SUBDUCTION

MODERN APPROACHES IN GEOPHYSICS

formerly *Seismology and Exploration Geophysics*

VOLUME 11

The titles published in this series are listed at the end of this volume.

SUBDUCTION
Insights from Physical Modeling

by

ALEXANDER I. SHEMENDA
Laboratoire de Géophysique et Tectonique,
Université de Montpellier II,
Montpellier, France

KLUWER ACADEMIC PUBLISHERS
DORDRECHT / BOSTON / LONDON

Library of Congress Cataloging-in-Publication Data

Shemenda, Alexander I.
Subduction : insights from physical modeling / by Alexander I. Shemenda.
p. cm. -- (Modern approaches in geophysics ; v. 11)
Includes bibliographical references and index.
ISBN 0-7923-3042-0 (acid-free)
1. Subduction zones. 2. Geological modeling. I. Title. II. Series.
QE511.46.S54 1994
551.1'36--dc20 94-30335

ISBN 0-7923-3042-0

Published by Kluwer Academic Publishers,
P.O. Box 17, 3300 AA Dordrecht, The Netherlands.

Kluwer Academic Publishers incorporates
the publishing programmes of
D. Reidel, Martinus Nijhoff, Dr W. Junk and MTP Press.

Sold and distributed in the U.S.A. and Canada
by Kluwer Academic Publishers,
101 Philip Drive, Norwell, MA 02061, U.S.A.

In all other countries, sold and distributed
by Kluwer Academic Publishers Group,
P.O. Box 322, 3300 AH Dordrecht, The Netherlands.

Printed on acid-free paper

Printed in the Netherlands

TABLE OF CONTENTS

PREFACE

Of the two modeling methods, mathematical (numerical) and physical (experimental), I would prefer dealing with a computer, leaving the "dirty" experimental work to those who enjoy working with real physical objects. Mathematical modeling enables us to more easily vary parameters over a wide range and create such conditions and models which can never be implemented in the laboratory. However, this method encounters serious problems in constructing ever more complex and adequate models of geologic processes in spite of the great possibilities provided by modern supercomputers. Frequently, it is not even clear how to formulate the problem mathematically, for example, in cases where it involves large transient deformations and failure of the lithosphere at the plate boundaries, in particular at the convergent boundaries. In many cases, physical modeling can in principle obviate these difficulties which makes this method more attractive and deserving of further development. The main problem is to put the experiments on a quantitative basis, i.e. to create models that satisfy the necessary requirements, the similarity criteria, in the first place.

This work represents a coherent description of the physical (mechanical) modeling of subduction and subduction-related phenomena, including the similarity criteria, the experimental technique, results of the model experiments, theoretical analysis of the results on the basis of continuum mechanics, and their geodynamic interpretation.

The problem of physical similarity is considered in the first chapter. To create a physically similar laboratory model of subduction we apply a novel technique developed on the basis of using hydrocarbon model materials for modeling the lithosphere with properties strongly dependent on temperature. This technique and experimental facilities are described in the second chapter. Experiments are performed with elastico-plastic lithospheric models underlain by a low-viscosity asthenosphere. We use both 2-D and 3-D approaches. Driving forces of subduction are both lateral compression of the lithosphere by the piston and gravitational sinking of the subducting plate. Different modes of initiation and development of subduction under horizontal compression are studied in the third chapter. Subduction is modeled in general as well as applied to particular regions such as the Central Basin of the Indian Ocean where a new subduction zone is proposed to be initiating.

Chapter 4 concentrates on the back arc dynamics and its relation to subduction regime. The effect of hydrostatic suction between plates in the subduction zone is shown to be responsible for the back arc tension. The effect is due to the low effective strength of the lithosphere compared with the hydrostatic stresses within it. The back arc stress regime correlates with a departure from isostatic equilibrium in the arc-trench area and, hence, can be recognized in the free air gravity anomaly field. Situations in some particular regions with active back arc basins are considered as well. The Aegean region where back arc tension can not be explained using traditional models is one of them.

Chapter 5 deals with situations where the lithosphere in a subduction zone is subjected to a high compression caused by subduction of very young lithosphere, different asperities and especially of the continental margin. Under strong compression the lithosphere can fail in a new place resulting in a jump of the subduction zone to that place. Location of the jump, style of the lithospheric deformation preceding jump, as well as polarity of a new subduction zone depend on particular conditions. For example, subduction of the Asian margin under the Luzon arc seems to result in initiation of new subduction under northeastern Taiwan.

Chapter 6 presents results of the modeling of continent-continent collision with three-layer lithospheric model which includes a very weak lower crust. The crust undergoes complex deformation in the collision zone, being scraped from the underlying mantle layer while this layer itself subducts similar to the subduction of the oceanic lithosphere in "classical" subduction zones. Such a similarity justifies the title of the chapter *Continental subduction* and its inclusion in this book.

ACKNOWLEDGEMENTS

The experimental results represented in this book were obtained mainly during work at the Institute of Oceanology, USSR Academy of Sciences and then at Moscow State University. I am grateful to L.I. Lobkovsky, O.G. Sorokhtin, and A.S. Ushakov who provided support with the organization of the physical modeling at that time and with whom some joint work has been done. Many of the experiments were performed together with A.L. Groholsky. I wish to express thanks to him and to E.P. Semenov for the long time cooperation and technical assistance. During different periods of work on the problem, I also had a fruitful cooperation and useful discussions with A.N. Bokun, E.P. Dubinin, Y.I. Galushkin, P.N. Kropotkin, S. Lallemand, J. Malaville, M. Mattauer, Y.I. Prozorov, A.L. Savostin, E.A. Tischenko, L.P. Zonenshain, and others. Thanks to them all. I am grateful to M.-A. Gutscher for critical improvement of the English. I would like to give my special thanks to S. Klessova for the manuscript preparation.

The experimental results in Section 5.5 on geodynamics of Taiwan were obtained together with R. K. Yang, C. H. Hsieh, and J. C. Tang with support by NSC of Taiwan and have been included in the book with their kind permission. The gravity map in Figure 5.17 was provided by G. C. Tsuei, the diagram in Figure 5.15 by C. H. Liu and the seismicity cross sections in Figure 5.16 by R. D. Huang. I am grateful to all of them and my numerous other Taiwanese colleagues for the useful discussions and help during my work in Taiwan where part of the book has been written.

CHAPTER 1

SIMILARITY CRITERIA

1.1. Introduction

Since the large spatial and time scales of geodynamic processes renders their direct observation and investigation impossible in most cases, the methods of quantitative modeling become very important. Frequently, modeling is the only way to get some ideas about geological phenomena. Most popular is mathematical modeling. Another method, physical (scale, laboratory or experimental) modeling is relatively seldom used, in spite of its obvious merits. There are several reasons for this. The most important seems to be the difficulty of making laboratory models that satisfy the necessary requirements, similarity criteria in the first place. Experimental models that are only qualitatively similar to the original no longer seem satisfactory to researchers and are below the level of mathematical modeling. At the same time the need for correctly set laboratory experiments is increasing. This is largely associated with the mathematical difficulties which numerical modeling encounters when constructing progressively more complex, and adequate models. Compact analytic or numerical solutions can frequently be obtained only by excessively simplifying the formulation of the problem. These simplifications largely determine the final result. Sometimes it is not even clear how to formulate the problem mathematically. Physical modeling can in principle obviate these difficulties. A major problem is the fulfillment of similarity criteria.

The question of physical similarity for experimental modeling involves two aspects, theoretical and experimental. The theoretical aspect consists in the derivation of the similarity criteria, while the experimental aspect proper consists in the finding or making model materials and experimental conditions that satisfy the similarity criteria. The main difficulties are associated with the creation of model materials and the implementation of the model process while the theoretical aspect seems to cause no major problems. The theory of physical similarity has been developed in the set of old general classical works [*Buckingham*, 1914; *Bridgman*, 1931; *Hubbert*, 1937; *Langhaar*, 1951; *Sedov*, 1959] and was considered in numerous more recent papers including those for geological applications [*Kline*, 1965; *Shimamoto*, 1974; *Gzovsky*, 1975; *Ramberg*, 1981; *Shemenda*, 1983; *Sherman*, 1984; *Weijermars and Schmeling*, 1986; *Ranalli*, 1987; *Davy and Cobbold*, 1991 etc.]. Similarity criteria can be derived in two ways. The first most general one is based on the dimensional analyses and the Pi-theorem. The criteria can be easily obtained by this method if the set of parameters governing (controlling) the investigated phenomenon is known. Boundary and initial conditions can be properly converted from prototype to model as well. In fact, it is only necessary to determine correctly the whole set of the controlling parameters. Strictly it can be done if the mathematical formulation of the problem is available, because the system of equations contains all the parameters. There is however the danger of selecting superfluous parameters, as some of them can be used in combination as one parameter. Such a danger can be eliminated using a standard procedure for the transformation of the equations into dimensionless form. This is the second way of deducing the similarity criteria. Unfortunately, frequently the whole set of the equations describing the phenomenon in question is not known and the only way to obtain the criteria is to combine both methods. This chapter presents a deduction of similarity criteria important for geodynamics (not only for subduction) using both methods. Some of the criteria are well known and widely used, while the others have been derived for the first time. We discuss some practical problems of physical similarity in laboratory modeling as well.

1.2. The General Approach to Physical Similarity

Suppose we have two geometrically similar objects, one, the original (prototype), has large dimensions and another, the model, is smaller. The properties of the model and the original are also similar, but the quantitative parameters characterizing the properties are different. For example, if the prototype possesses linear-viscous properties, then the model has also such properties, but the coefficient of viscosity may be different. Boundary and initial conditions are also similar (boundary forces act in the same directions, but their values may be different and so on). Using the language of mathematics, one can say that we are dealing with phenomena that are described by systems of identical equations, including boundary and initial conditions. We suppose that the range of parameter values is such that the solutions for the model have the same form as for the original. The task is to find relations (similarity criteria) between the parameters for the model and the prototype which being satisfied ensure physical (not simply geometrical) similarity between two objects. Physical similarity means that any variable characterizing appropriate points in the model and original at the appropriate times in the model and original are proportional according to a calculable coefficient. For a given variable (for example, rate vector or deformation) this coefficient is the same for all corresponding points and times in the model and original. This brief and somewhat vague definition is clarified below.

Let $y^m = f(z_i^m)$ and $y^o = f(z_i^o)$ be relevant solutions of the equation systems for the model and the original (say, y is one of stress-tensor components) where f is a function; z_i are all constant and variable arguments that characterize (control) the phenomenon under consideration; the superscripts "o" and "m" here and below stand for original and model, respectively; i = 1, 2,... n. The sets of controlling parameters z_i for the model and original are assumed to be identical, only the values of corresponding parameters z_i^m and z_i^o being different, that is,

$$
\begin{gathered}
z_1^m = \mu_1 z_1^o \\
z_2^m = \mu_2 z_2^o \\
\cdots\cdots\cdots \\
z_n^m = \mu_n z_n^o
\end{gathered}
$$

or

$$z_i^m = \mu_i z_i^o \ (i = 1, 2 \ldots n), \tag{1.1}$$

where the μ_i are some constants (similarity coefficients). The condition for physical similarity between original and model can be now written as

$$f(z_i^m) = \varphi(\mu_i) \times f(z_i^o) \tag{1.2}$$

where $\varphi(\mu_i)$ is a function which for given f and μ_i (i = 1, 2 ... n) is a constant. Thus, the determination of similarity criteria reduces to finding a transformation of the form (1.1) for which (1.2) holds under any f. The desired transformation follows from the Pi-theorem [*Buckingham*, 1914] and can be expressed as [*Sedov*, 1972]

$$z_i^m = \mu_i \times z_i^o \ (i \le k); \ z_i^m = \mu_1^{q_1^{i-k}} \times \mu_2^{q_2^{i-k}} \times \ldots \mu_k^{q_k^{i-k}} \times z_i^o \ (i > k) \tag{1.3}$$

Here, z_i ($i \le k$) are quantities with independent dimensions; μ_i ($i \le k$) are transformation coefficients for the k parameters from (1.1) with independent dimensions; k is the number of controlling parameters with independent dimensions; the numbers q_j^{i-k} (i and j are indices; $i > k$; $j = 1, 2, \ldots k$) are obtained from

$$[z_i] = Q_1^{q_1^{i-k}} \times Q_2^{q_2^{i-k}} \times \ldots Q_k^{q_k^{i-k}} \quad (i > k), \tag{1.4}$$

where the notation [ ] stands for the dimension of the relevant parameter; $Q_i = [z_i]$ ($i \le k$). The function $y(z_i)$ itself is transformed as $y^m = \mu_1^{a_1} \times \mu_2^{a_2} \times \ldots \mu_k^{a_k} \times y^o$, where the a_i ($i \le k$) are found from $[y] = Q_1^{a_1} \times Q_2^{a_2} \times \ldots Q_k^{a_k}$.

Thus if the phenomenon we wish to study is defined by n parameters, k having independent dimensions, then k parameters can be selected arbitrarily, the remaining n-k ones being uniquely determined by this selected set (see (1.3)).

Transformation coefficients μ_i for dimensionless parameters must equal unity, i.e., the similarity condition demands that dimensionless parameters (Poisson's ratio, coefficients of volumetric and linear expansion, angles and so on) should be equal in the model and the original. To see this, let $z_{k+\gamma}$ ($\gamma \le n-k$) be a dimensionless parameter. Recalling that the Q_i ($i \le k$) are independent dimensions, we deduce from (1.4) that all q_i^{γ} ($i \le k$) must vanish. It then follows from (1.3) that $\mu_{k+\gamma} = 1$.

Eliminating μ_i from (1.3), we get relations which must be satisfied for the model to be similar to the original

$$\frac{z_i^m}{z_i^o} = \left(\frac{z_1^m}{z_1^o}\right)^{q_1^{i-k}} \times \left(\frac{z_2^m}{z_2^o}\right)^{q_2^{i-k}} \times \ldots \left(\frac{z_k^m}{z_k^o}\right)^{q_k^{i-k}}. \tag{1.5}$$

Function y is converted to the original as follows

$$y^{o} = \left(\frac{z_1^{o}}{z_1^{m}}\right)^{a_1} \times \left(\frac{z_2^{o}}{z_2^{m}}\right)^{a_2} \times \cdots \left(\frac{z_k^{o}}{z_k^{m}}\right)^{a_k} \times y^{m}. \tag{1.6}$$

Conditions (1.5) can be rewritten in a different form in which they are usually called similarity criteria

$$z_i/(z_1^{q_1^{i-k}} \times z_2^{q_2^{i-k}} \times \ldots z_k^{q_k^{i-k}}) = C_{i-k} \text{ (i > k)},$$

or when written out

$$\begin{aligned} z_{k+1}/(z_1^{q_1^1} \times z_2^{q_2^1} \times \ldots z_k^{q_k^1}) &= C_1 \\ z_{k+2}/(z_1^{q_1^2} \times z_2^{q_2^2} \times \ldots z_k^{q_k^2}) &= C_2 \\ \cdot \quad \cdot \quad \cdot \quad \cdot \quad \cdot \quad \cdot \quad \cdot \quad \cdot \quad \cdot \\ z_n/(z_1^{q_1^{n-k}} \times z_2^{q_2^{n-k}} \times \ldots z_k^{q_k^{n-k}}) &= C_{n-k} \end{aligned} \tag{1.7}$$

where the $C_{1,2...}$ are constants which must be identical when substituting the parameters for the model and the original.

To summarize, if a set of controlling parameters is defined, relations (1.7) yield conditions which, when being satisfied, ensure physical similarity between the model and prototype (boundary and initial conditions should be transformed using (1.6)).

Consider the following example. Let the controlling parameters of a tectonic process be η, H, V, t, ρg which are the viscosity coefficient, the linear size typical of the process (layer thickness, say), the velocity, the time, and the specific weight (the density times the acceleration of gravity), respectively. This set does not include Cartesian coordinates, as they would merely yield the trivial conditions for geometrical similarity. The SI dimensions of the parameters are: $[\eta]$ = kg/m × s; $[V]$ = m/s; $[H]$ = m; $[t]$ = s; $[\rho g]$ = kg/m^{2}s^2. One can easily see that only three parameters have independent dimensions (k = 3). If we select η, V, and H as independent (denoting them z_1, z_2, z_3) then the dimensions for the remaining two parameters, t (z_4) and ρg (z_5), are expressible in terms of these: $[t] = [V]^{-1} \times [H]^{1}$; $[\rho g] = [\eta]^{1} \times [V]^{1} \times [H]^{-2}$. According to (1.4) one has $q_1^1 = 0$; $q_2^1 = -1$; $q_3^1 = 1$; $q_1^2 = 1$; $q_2^2 = 1$; $q_3^2 = -2$. Substitution of these values into (1.7) yields n - k = 2 similarity criteria

$$Vt/H = \text{constant}; \qquad \eta V/(\rho g H^2) = \text{constant}. \tag{1.8}$$

All combinations of these criteria, whether multiplications, ratios, or raising to a power, can also be regarded as similarity criteria. They will evidently be automatically satisfied, provided that independent similarity criteria (1.8) are themselves satisfied (any two conditions in terms of which (1.8) are expressible can be taken as independent).

Suppose a boundary condition in the original is a force F^{o}, say, acting across a boundary. The corresponding force in the model F^{m} is determined as follows: The dimension of force is expressed in terms of the independent dimensions of η, V, H: $[F] = \mathrm{kg} \times \mathrm{m/s^2} = [\eta]^1 \times [V]^1 \times [H]^1$, and then the use of (1.6) with F substituted for y yields the desired expression

$$F^{\mathrm{o}} = \left(\frac{\eta^{\mathrm{o}}}{\eta^{\mathrm{m}}}\right)^1 \times \left(\frac{V^{\mathrm{o}}}{V^{\mathrm{m}}}\right)^1 \times \left(\frac{H^{\mathrm{o}}}{H^{\mathrm{m}}}\right)^1 \times F^{\mathrm{m}} \qquad \text{or} \quad F^{\mathrm{m}} = \frac{\eta^{\mathrm{m}} V^{\mathrm{m}} H^{\mathrm{m}}}{\eta^{\mathrm{o}} V^{\mathrm{o}} H^{\mathrm{o}}} \times F^{\mathrm{o}}.$$

1.3. Derivation of Similarity Criteria from the Equations of Continuum Mechanics

Previous considerations show that there is no problem with similarity criteria if a set of controlling parameters is specified, but the question is how to determine these parameters correctly. Strictly, it can be done only within the framework of a specific mathematical model of the phenomenon which incorporates all essential characteristics and driving forces. Since tectonics is primarily concerned with mechanical processes, we will deal with models of continuum mechanics. A system of equations describing mechanical processes includes equations of motion (or of equilibrium), an equation of continuity, an equation of conservation of energy, equations relating stress, strain, strain rate, temperature, pressure and other physical and chemical parameters of state. Along with these equations, the description of a process includes initial and boundary conditions.

Common for all processes are the following equations of motion

$$\rho a_{\mathrm{j}} = \rho F_{\mathrm{j}} + \partial \sigma_{\mathrm{ij}} / \partial x_{\mathrm{i}} \tag{1.9}$$

and an equation of continuity written below for an incompressible homogeneous medium

$$\dot{\varepsilon}_{ij}\delta_{ij} = 0; \tag{1.10}$$

here

$$a_j = \partial v_j/\partial t + v_i \partial v_j/\partial x_i \tag{1.11}$$

are components of acceleration **a** (**a** = d**V**/d*t*);

$$\dot{\varepsilon}_{ij} = -\frac{1}{2}(\partial v_i/\partial x_j + \partial v_j/\partial x_i) \tag{1.12}$$

is strain rate tensor; σ_{ij} is stress tensor; v_i are components of the velocity vector **V**; ρ is the density; F_j are components of the body force; t is the time; δ_{ij} is the Kronecker's delta; x_i are Cartesian coordinates; the subscripts i, j take on the values 1, 2, 3.

The specific form for the equation of conservation of energy which is the first law of thermodynamics, the equation of thermodynamic state, and the equations of rheologic state are based on experimental data for the particular medium.

If the properties of the medium are little affected by thermodynamic parameters (in particular, temperature), which is frequently the case in practice, then the equation of state and the law of conservation of energy may be excluded from the general system of equations. As to the equations describing the properties of the medium, that the simplest and widely used ones in geodynamics are the Navier-Stokes, the Mises's and the Hooke's equations which describe a viscous (linearly viscous), plastic, and elastic models, respectively. Below we consider these models in more detail.

1.3.1. The Linearly Viscous Fluid

The motion of linearly viscous incompressible fluid (with zero coefficient of volumetric viscosity) is described by equations (1.9), (1.10), the Navier-Stokes law

$$\sigma_{ij} = \eta\dot{\varepsilon}_{ij} - p\delta_{ij} \tag{1.13}$$

and the appropriate boundary and initial conditions. Here, p is pressure and η is the coefficient of dynamic (shear) viscosity. Equations (1.9) to (1.13) contain all parameters that are essential for the dynamics of viscous fluid; they are η, ρ, V, t, H and g, where H is the scale of x_i; g the acceleration of gravity (the F_j in (1.9) equals ρg along the vertical axis; $F_j = 0$ along the other axes). Only three of these parameters have independent dimensions. Therefore one can create three dimensionless expressions (see (1.7))

$$\rho VH/\eta = \text{const}\,;\quad Vt/H = \text{const}\,;\quad \eta V/(\rho g H^2) = \text{const}. \tag{1.14}$$

(the first and second expressions are Reynolds and Struchal numbers). The same conditions can also be derived directly from equations (1.9) to (1.13). To do this, we make the change of variables: $x_i = H\tilde{x}_i$; $v_i = V\tilde{v}_i$; $t = \tau\tilde{t}$; $\sigma_{ij} = \rho g H\tilde{\sigma}_{ij}$; $p = \rho g H\tilde{p}$. Here, the dimensionless variables marked with tildes are preceded by scales of the corresponding dimensional values for given phenomenon. Substituting these expressions and relations (1.11), (1.12) into (1.9), (1.13) and replacing τ with t, we get

$$\frac{\rho V}{t}\times\frac{\partial\tilde{v}_i}{\partial\tilde{t}} + \frac{\rho V^2}{H}\tilde{v}_i\times\partial\tilde{v}_j/\partial\tilde{x}_i = \rho g + \rho g\times\partial\tilde{\sigma}_{ij}/\partial\tilde{x}_i \tag{1.15}$$

$$\rho g H\tilde{\sigma}_{ij} = \frac{1}{2}\frac{\eta V}{H}\times(\partial\tilde{v}_i/\partial\tilde{x}_j + \partial\tilde{v}_j/\partial\tilde{x}_i) - \rho g H\tilde{p}\delta_{ij}. \tag{1.16}$$

Dividing (1.15) by ρg and (1.16) by $\rho g H$ yields

$$\frac{V}{gt}\times\partial\tilde{v}_j/\partial\tilde{t} + \frac{V^2}{gH}\tilde{v}_i\times\partial\tilde{v}_j/\partial\tilde{x}_i = 1 + \partial\tilde{\sigma}_{ij}/\partial\tilde{x}_i \tag{1.17}$$

$$\tilde{\sigma}_{ij} = -\tilde{p}\delta_{ij} + \frac{1}{2}\frac{\eta V}{\rho g H^2}\times(\partial\tilde{v}_i/\partial\tilde{x}_j + \partial\tilde{v}_j/\partial\tilde{x}_i) \tag{1.18}$$

These equations contain dimensionless coefficients (composed of constant parameters and scales of variables), dimensionless variables, and their derivatives. If these coefficients are equal for phenomena that differ in their geometrical and "physical" scales, then the systems of equations describing this phenomena will be identical. Thus, the equality of the coefficients $V/(gt)$; $V^2/(gH)$; $\eta V/(\rho g H^2)$ from (1.17) and (1.18) (equation (1.10) does not yield dimensionless expressions) for the model and the original provides their physical similarity. These coefficients can be easily transformed to the form (1.14).

It should be pointed out that conditions (1.14) are not all similarity criteria for the modeling of the motion of a linearly viscous fluid. To see this, consider the free motion of a fluid in the gravitational field, for example, the gravitational spreading of a topographic rise, the rise of a diapir etc. V and t in such processes are not controlling parameters but desired ones. For this reason they should be eliminated from (1.14). There remains a single similarity criterion

$$\eta^2/(\rho^2 g H^3) = \text{const} \tag{1.19}$$

which involves all the controlling parameters for the mentioned processes. When this criterion and the appropriately transformed initial and boundary conditions are satisfied, the model time t can be converted to nature using the first two conditions from (1.14)

$$t^o = [g^m H^o/(g^o H^m)]^{1/2} \times t^m, \tag{1.20}$$

while the velocities V^m and V^o at corresponding instants are related by

$$V^o = [g^o H^o/(g^m H^m)]^{1/2} \times V^m. \tag{1.21}$$

Now we estimate some of the model parameters. For the original we assume typical values $\eta^o = 10^{19}$ - 10^{21} Pa × s; $H^o = 10$ km; $\rho^o = 3 \times 10^3$ kg/m^3. For the model we take $\rho^m = 10^3$ g/m^3; $H^m = 10^{-2}$ m; $g^m = g^o = g$; we then use (1.19) to get $\eta^m = 10^9$ - 10^{11} Pa × s. This is a very high viscosity which nevertheless is attainable in models. Let the free process we study lasts $t^o = 10^{14}$ s (about 30 Ma) in nature, and its rate be $V^o = 10^{-9}$ m/s (≈ 3 cm/yr). The corresponding model values must be, according to (1.20) and (1.21) (when $g^m = g^o$): $t^m = 10^{11}$ s (≈ 3000 years); $V^m = 10^{-12}$ m/s (3×10^{-3} cm/yr). These are obviously unacceptable values for modeling in the laboratory. One can reduce the time t^m without violating the similarity conditions by using centrifuged models, i.e., by increasing g^m. Actually attainable values of g^m are of the order of (10^3 - 10^4)g. This diminishes t^m by two orders only, i.e., t^m can be as low as 10^9s or 30 years which still is unrealistic for the experiment. Varying the other model parameters within reasonable limits does not yield an acceptable result either. Hence one might conclude that tectonic processes cannot be adequately modeled in the laboratory. In fact, however, these difficulties do not arise, because the inertia terms (the left-hand sides of motion equations (1.9)) for slow tectonic processes are negligibly small. When these terms are discarded, the first two conditions in (1.14), Reynolds and Struchal numbers, vanish, leaving the third only. This is a similarity criterion for forced fluid motions, when the process velocity V is fixed (i.e., when V is the controlling parameter). The remaining condition $\eta V/(\rho g H^2)$ = constant means that the ratio of viscous stress $\eta V/H$ to hydrostatic pressure $\rho g H$ for the model and the original should be equal. There are no principal problems in fulfilling this criterion.

Along with the inertia terms, the time t is also eliminated from equations describing the motion of a viscous fluid. That does not mean that we are only considering a steady state (time-independent) motion of a viscous fluid. The nonstationary nature of a tectonic process is determined by boundary conditions or by some other (non-inertial) physical mechanisms. For example, the nonstationarity of the above-mentioned flow of a topographic rise in the gravitational field is due to a time-dependent free boundary (the surface of the rise) and is incorporated into a kinematic condition at the moving boundary

$$\partial S/\partial t + v_i \partial S/\partial x_i = 0, \tag{1.22}$$

where $S(x_i,t) = 0$ is the equation of the free surface.

If the motion of a fluid is due to its chemical inhomogeneity (concentrational convection, diapirism), then the description of the motion incorporates the time variable into the equation of continuity which can now be written in a more general form compared with (1.10)

$$\partial \rho/\partial t + \partial(\rho v_i)/\partial x_i = 0. \tag{1.23}$$

We again get the criterion (Struchal number)

$$Vt/H = \text{const} \tag{1.24}$$

from (1.22) and (1.23) which is responsible for the transient character of the motion. The physical meaning of this condition is that the ratio of current time t to the time scale H/V that is typical of this transient process must be the same in the model and the original. The condition (1.24) for steady state processes acquires a purely kinematic (geometrical) sense, namely, the ratio of path passed (displacement) Vt to the typical linear dimension H should be the same in the model and the original. The slow motion of a viscous fluid is thus characterized by two similarity criteria

$$Vt/H = \text{const}\ ;\ \eta V/(\rho g H^2) = \text{const} \tag{1.25}$$

(the Reynolds number should be added to these when modeling rapid processes). Note that (1.25) are identical with (1.8) which were deduced from (1.7) by replacing the two controlling parameters ρ and g by a single one ρg.

The expressions (1.25) are not similarity criteria in modeling free motion processes like those considered above (the gravitational spreading of a topographic rise, the rise of a diapir etc.). Such processes can be modeled with any values of η^m, ρ^m, H^m, g^m, provided that inertia forces in the model are small compared with viscous frictional forces, or (which is the same) the Reynolds number is small compared with unity. The velocity V^m and the time t^m are not fixed in modeling free motion. They are functions of η^m, ρ^m, H^m, g^m, are measured experimentally, and then converted to nature using (1.25):

$$V^o = \frac{\rho^o H^{o2} \eta^m g^o}{\rho^m H^{m2} \eta^o g^m} \times V^m,$$

$$t^o = \frac{\rho^m H^m \eta^o g^m}{\rho^o H^o \eta^m g^o} \times t^m. \tag{1.26}$$

When, however, a forced motion is modeled, or a process where a forced motion with velocity V_0^o in nature is superposed on a free motion, the corresponding model parameter V_0^m may no longer be arbitrary, but is given by the second condition in (1.25) (see (1.26)) which becomes a similarity criterion in that case.

Convective motion. Consider the slow free motion of a viscous incompressible fluid driven by heating. This is thermal convection which can be described by adding to equations (1.9) - (1.13) the equation for heat transfer (the equation of heat conduction in a moving medium)

$$\partial T/\partial t + v_i\, \partial T/\partial x_i = \kappa\, \Delta T \tag{1.27}$$

(this form does not incorporate heat release due to viscous dissipation) and the equation of state [*Gershuni and Zhukhovitsky*, 1972]

$$\rho = \rho_o(1 - \beta_V T), \tag{1.28}$$

where T is the departure of temperature from the equilibrium distribution; ρ_o is the density at $T = 0$; κ is the thermal diffusivity: $\kappa = \lambda/\rho C_p$; C_p is specific heat at constant pressure; λ is thermal conductivity; β_V is the coefficient of thermal expansion. Converting (1.27) and (1.28) to dimensionless form as was done for equations (1.9) to (1.13), we get

$$\frac{H^2}{\kappa t} = \text{const}; \quad \frac{VH}{\kappa} = \text{const}; \quad \beta_V \Delta T = \text{const},$$

where ΔT is the typical temperature contrast in the system (T scale), usually the difference between the bottom and the top of a convective layer. Combining these conditions with (1.25) and eliminating the desired (non-governing) parameters V and t, we get similarity criteria for the modeling slow thermal convection

$$\frac{\rho g H^3}{\kappa \eta} = \text{const}; \quad \beta_V \Delta T = \text{const}. \tag{1.29}$$

These criteria can be reduced to a single one in the following way: Equation (1.28) relates the variation of density $\Delta\rho$ with that of temperature. The main effect of density variation is its influence on convection driving force, the buoyant force $\Delta\rho g = \pm\rho g \beta_V \Delta T$ which affects heated or cooled volumes of material. Therefore one can incorporate (1.28) only into the gravitational term in (1.9), treating the density constant in the other terms of equations (1.9) and (1.27). This is the Boussinesque's approximation which is in wide use for solving convection problems. One can easily see that the two criteria (1.29) are replaced in this approximation by a single one which is a product of criteria (1.29)

$$\frac{\rho \beta_V \Delta T g H^3}{\eta \kappa} = \text{const}. \tag{1.30}$$

This is the Rayleigh number. Another important parameter in the theory of thermal convection is the Prandtl number ($\eta/\rho\kappa$ = constant). It is not present in (1.29), because the derivation of these criteria was based on relations (1.25), i.e., inertia forces have been neglected.

Temperature on the boundaries of the convecting layer as well as viscosity η are not necessarily constants. For example, viscosity can depend on temperature, $\eta^o = \eta^o(T^o)$. This function is converted into the model using (1.30)

$$\eta^m(T^m) = \frac{\rho^m \beta_v{}^m \Delta T^m g^m H^{m3} \kappa^o}{\rho^o \beta_v{}^o \Delta T^o g^o H^{o3} \kappa^m} \times \eta^o(T^o)$$

where $T^m = T^o \Delta T^m / \Delta T^o$ (it is assumed for simplicity that the temperature at the upper boundary is zero for both model and prototype).

1.3.2. Phase Transition

Processes in which deformation of the media develops on the background of phase transition are described by the equations of motion for the relevant media and the equation for heat conduction with specific heat C_p' in the temperature range $T_s \leq T \leq T_l$ expressed as [*Ushakov et al.*, 1979]

$$C_p' = C_p + L/(T_l - T_s),$$

where T_s, T_l are solidus and liquidus temperatures, respectively; C_p is the specific heat capacity for $T > T_l$ and $T < T_s$; L is the latent heat. This equation yields an additional similarity criterion

$$\frac{L}{C_p(T_l - T_s)} = \text{const}$$

which should be used along with $H^2/\kappa t$ = const and (in case of modeling thermo-mechanical processes) with relevant mechanical criteria.

1.3.3. The Rigidly Plastic Body

Deformation of perfectly plastic material is described by equations (1.9) (without the left-hand side), (1.10), and the Mises' equations

$$s_{ij} = \frac{2\tau_s}{H_{\dot{e}}}\dot{e}_{ij}, \tag{1.31}$$

where $H_{\dot{e}} = (\frac{1}{2}\dot{e}_{ij}\dot{e}_{ij}))^{1/2}$ is the intensity of shear strain rate; s_{ij}, $\dot{e}_{ij}$ are deviatoric stress and strain rate tensors: $s_{ij} = \sigma_{ij} - \frac{1}{3}\sigma_{ij}\delta_{ij}$; $\dot{e}_{ij} = \dot{\varepsilon}_{ij} - \frac{1}{3}\dot{\varepsilon}_{ij}\delta_{ij}$; τ_s is the shear yield limit [*Kachanov*, 1969].

Deformation (flow) of the above considered viscous body begins at arbitrarily small stresses. In contrast to this, deformation in a plastic body starts only after the stress has exceeded a certain value, the yield limit, or to be more exact, when the Mises-Hencky yield condition has been satisfied

$$(\tfrac{1}{2}s_{ij}s_{ij})^{1/2} = \tau_s.$$

The equations (1.31) involve the strain rate $\dot{e}$ (hence the velocity *V*), however, in fact they only establish the relation between stress and displacement (velocity is used as displacement per unit time). Velocity and time do not actually enter the system of equations. Therefore, the problem with duration and rate of model process does not occur at all when the modeled object possesses plastic (rigidly (perfectly) plastic or elastico-plastic) properties. The analog of time in this case is the displacement Δl. If it is known, for instance, that the shortening of the investigated object in nature is Δl^o, then the corresponding value in the model should be $\Delta l^m = (H^m/H^o)\Delta l^o$, and it does not matter what is the velocity and duration of the shortening.

Equations (1.9, 1.10, 1.31) which describe the behaviour of a rigidly plastic body yields a single similarity criterion

$$\tau_s/\rho g H = \text{const}, \tag{1.32}$$

which means that the ratio of plastic stress to hydrostatic (lithostatic) pressure should be the same for the model and original.

1.3.4. The Linearly Elastic Body

Stress-strain state of such a body is described by equations of equilibrium and Hooke's equations

$$\sigma_{ij} = \frac{E\nu}{(1+\nu)(1-2\nu)} I_1(\varepsilon)\delta_{ij} + \frac{E}{1-\nu}\varepsilon_{ij}, \qquad (1.33)$$

where ε_{ij} is strain tensor; $I_1(\varepsilon) = \varepsilon_{ij}\delta_{ij}$ is the first invariant of the strain tensor; E and ν are the modulus of elasticity and Poisson's ratio. The equations give two similarity criteria

$$E/\rho g H = \text{const}; \quad \nu = \text{const}, \qquad (1.34)$$

which should be satisfied in modeling of small elastic deformation. The last criterion in (1.34) can frequently be discarded, as in many cases the Poisson's ratio does not significantly affect the result.

1.3.5. Complex Rheologic Models

We considered above the simplest (elementary) models. Greater accuracy (or adequacy) in description of the properties of real tectonic media can be achieved by combination of these models (elements). Any complex model constructed of the elementary models (see Figure 1.1) can be described by equations (1.9) to (1.13), (1.31), and (1.33) and some extra conditions determined by the way of connection of the elements. For example, for Maxwell's elastico-viscous model (Figure 1.1*a*) we have

$$\dot{\varepsilon}_{ij} = \dot{\varepsilon}_{ij}^{e} + \dot{\varepsilon}_{ij}^{v}; \quad \sigma_{ij} = \sigma_{ij}^{e} = \sigma_{ij}^{v} \qquad (1.35)$$

where the superscripts "e" and "v" refer to elastic and viscous elements, respectively; σ_{ij}, $\dot{\varepsilon}_{ij}$ are the resulting stress and strain rates for the whole combination of elements. Relations like (1.35) do not provide similarity criteria. For this reason, independently of the way the elements are combined, the similarity criteria for any complex rheologic model will be all

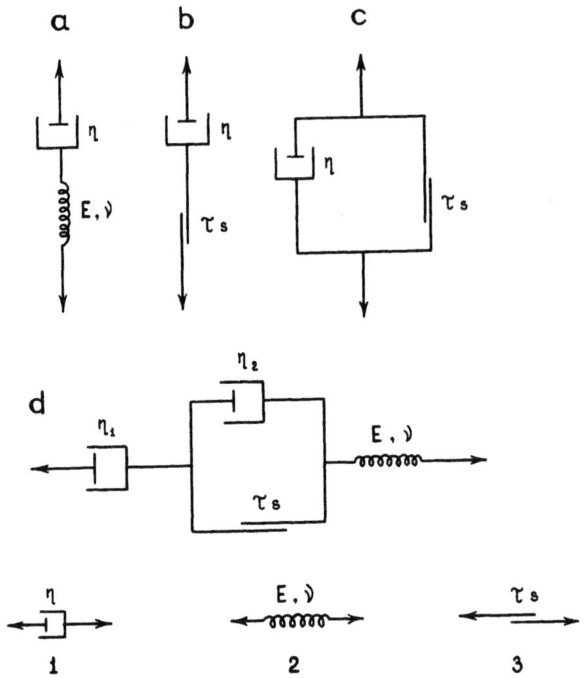

Figure 1.1. Complex rheologic models
(*a*) Maxwell's visco-elastic model; (*b*) and (*c*) Visco-plastic models with different combinations of elements; (*d*) Model approximating the properties of clay [*Gzovsky*, 1975];
1, 2, 3 are the viscous, elastic, and plastic elements, respectively.

of the criteria for the component elements taken together. In particular, the similarity criteria for the elastico-plastic models are (1.32) and (1.34).

The way the elements are combined should be taken into account when selecting model materials. It is not sufficient, for instance, that the material designed to imitate a visco-plastic medium have visco-plastic properties. It is necessary that viscous and plastic properties obey certain qualitative and quantitative relations, the same ones as in the original. The way rheologic elements are combined also determines the technique for measurement of the parameters η, τ_s, and E. For example, in the case shown in Figure 1.1*b*, to determine η it is necessary to apply to the sample a stress τ less than τ_s. The value of η will then be $\eta = \tau/\dot{\varepsilon}$ where $\dot{\varepsilon}$ is the strain rate at stress τ. The yield limit τ_s can be determined at large $\dot{\varepsilon}$: $\dot{\varepsilon} > \tau_s/\eta$, i.e., when the maximum stresses τ in the system are equal to τ_s. For the model shown in Figure 1.1*c*, τ_s is determined at low $\dot{\varepsilon}$, when $\tau_s >> \eta\dot{\varepsilon}$, while η is measured at high strain rates ($\dot{\varepsilon} >> \tau_s/\eta$), when the yield limit can be neglected compared with the viscous stress $\eta\dot{\varepsilon}$.

One of the most popular model material for tectonic modeling is clay. On the basis of the elastic-visco-plastic models considered, clay can be represented by the scheme in Figure 1.1*d*. According to this scheme at large $\dot{\varepsilon}$ the deformation of the clay can mainly develop due to the elastic element, because viscous deformation at large $\dot{\varepsilon}$ requires large stresses. Therefore clay behaves as an elastic body under rapid (instantaneous) loading. If clay is under constant stress τ ($\tau < \tau_s$) or is being loaded slowly enough, then creep deformation due to the viscous element (η_1) develops. This deformation may be considerably greater than elastic deformation. When $\tau > \tau_s$, the elements (τ_s) and (η_2) begin to operate, in addition to the viscous (η_1) and elastic elements. In this regime the effective clay viscosity is η_2 which is considerably lower then η_1 (the viscosity at $\tau < \tau_s$), even though the stress equal to the sum $\tau = \tau_s + \eta_2\dot{\varepsilon}$ remains large. The rheologic properties of clay are thus controlled by four parameters η_1, η_2, τ_s, E (see Figure 1.1*d*), and each should satisfy similarity criteria. Usually such a requirement cannot be fulfilled in practice. Similarity can be achieved for one or at most two parameters. Sometimes this is enough to create more or less adequate model, sometimes it is not. In fact, the situation is even more bad because this comparatively complicated model (Figure 1.1*d*) is still a rather rough approximation of the properties of tectonic media, as well as clay itself. Experiments with rocks under different *P-T* conditions show that the $\tau(\dot{\varepsilon})$ function is essentially nonlinear and is "weaker" than in the Newtonian model due to the dislocation mechanism responsible for deformation in rocks [*Kirby*, 1983]. This function has the form

$$\dot{\varepsilon}_{ij} = A\tau_{ij}^{n}\,; \tag{1.36}$$

where A is constant for given P and T.

A weak dependence of τ on $\dot{\varepsilon}$ is also typical in a certain range of $\dot{\varepsilon}$ of many model materials which are structured tixotropic colloidal systems, one of these being clay. The nonlinear relationship $\tau(\dot{\varepsilon})$ for such systems is due to a deformation mechanism which is quite different from that in rocks [*Shemenda*, 1981, 1984]. The mechanism is associated with destruction and recovery of molecular bonds in the structural frame of the system during deformation. There is a certain rate of bond destruction for every value of $\dot{\varepsilon}$. When $\dot{\varepsilon}$ is small enough, all bonds destroyed by deformation may have time to recover. In that case the behaviour of the material is perfectly plastic. Beginning from some $\dot{\varepsilon}$ the bonds destroyed fail to recover fully. This diminishes $d\tau/d\dot{\varepsilon}$, so that each next increase in $\dot{\varepsilon}$ requires progressively smaller increments of τ (the derivative $d\tau/d\dot{\varepsilon}$ in the Newtonian model is a constant and is equal to the coefficient of viscosity η). Hence the properties of model materials can be even closer to the properties of real rocks than to the simple linear Newtonian model which is traditionally used to compare tectonic and model media.

If (1.36) is adopted instead of Navier-Stokes law, equations (1.16) are replaced by

$$\rho g H \tilde{\sigma}_{ij} = (V/AH)^{1/n} \times (\partial \tilde{v}_i/\partial \tilde{x}_j + \partial \tilde{v}_j/\partial \tilde{x}_i) - \rho g H \tilde{p}.$$

Two similarity criteria for modeling forced processes follow from this equation [*Shemenda*, 1987]

$$\frac{1}{\rho g H} \times \left(\frac{V}{AH}\right)^{1/n} = \text{const}; \quad n = \text{const} \tag{1.37}$$

(they replace the last criterion in (1.25)).

For a free motion there remains a single criterion, the last in (1.37). The modeling of thermal convection requires an additional criterion (an analog of the Rayleigh number)

$$\frac{A\beta_v \Delta T H^2}{\kappa} \times (\rho g H)^n = \text{const}.$$

The relationship between time and rate in the model and the original for the law (1.36) can be derived using the Struchal number and the first criterion from (1.37)

$$t^o = \frac{A^m}{A^o}\left(\frac{\rho^m g^m H^m}{\rho^o g^o H^o}\right)^n \times t^m.$$

1.4. Dilatant Plasticity and Brittle Failure

As was indicated in previous section the deformation of structured systems (model materials) is accompanied by failure of molecular bonds in the structural frame. Such a failure is manifested in the appearance, growth and branching of microcracks within the deforming specimen. This effect becomes especially conspicuous when τ approaches the yield limit τ_s and the $\tau(\varepsilon)$ curve in Figure 1.2 becomes subhorizontal, i.e., when the parallel elements (η_2) and (τ_s) in Figure 1.1*d* start to operate. The general pattern of deformation, in particular the direction of cracking, is controlled by dilatant properties of the material and pressure-dependent internal friction. The dilatancy and internal friction are also important for rocks that are in a dilatant plastic state typical of the upper lithosphere where the lithostatic pressure is not very great [*Nikolaevsky*, 1983]. These properties are incorporated into a more sophisticated theory in which plastic (inelastic) strain increments are determined as [*Rudnicki and Rice*, 1975]

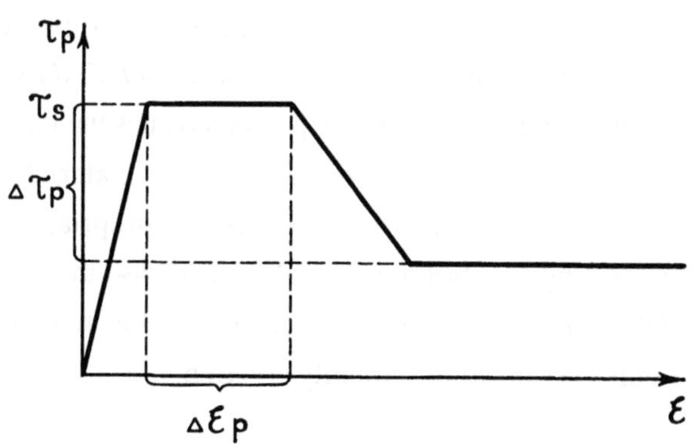

Figure 1.2. A simplified stress-strain curve for rocks and colloidal model materials.

$$2de_{ij} = \frac{s_{ij}}{h\bar{\tau}}[d\bar{\tau} + \mu\frac{d\sigma_{kk}}{3}] , \qquad (1.38)$$

$$d\varepsilon_{ij} = \beta(2de_{ij}de_{ij})^{1/2} , \qquad (1.39)$$

where $\bar{\tau} = (\frac{1}{2}s_{ij}s_{ij})^{1/2}$ is intensity of deviatoric stresses; β, μ and h are the dilatancy factor, internal friction coefficient, and plastic hardening modulus, respectively. These equations yield three more similarity criteria in addition to (1.32)

$$\beta = \text{const}; \;\; \mu = \text{const}; \;\; h/\tau_s = \text{const}. \qquad (1.40)$$

The properties of real materials are characterized by another important parameter which is not included in the model, the magnitude of plastic strain prior to failure $\Delta\varepsilon_p$ (Figure 1.2) (failure in different materials occurs upon reaching different values of $\Delta\varepsilon_p$). Corresponding to this controlling parameter is the similarity criterion

$$\Delta\varepsilon_p = \text{const}.$$

The mathematical model (1.38, 1.39) does not explicitly contain parameters controlling the development (propagation) of cracks in the material, the Griffiths surface energy δG or critical coefficients of stress intensity K_c [*Cherepanov and Ershov*, 1977]. These are implicitly contained in the integral characteristic β which relates the intensity of shear strain to volumetric dilatation (expansion of pore space in the material). Such an integral description of material properties becomes evidently inappropriate for investigation of individual cracks propagating in a brittle material. In that case the controlling parameter should be the critical coefficient of stress intensity for the relevant type (mode) of deformation which yields the following similarity criterion

$$K_c/(\tau_s H^{1/2}) = \text{const}.$$

The strength contrast $\Delta\tau_p$ (Figure 1.2) before and after failure (which can be considered as a characteristic of material brittleness) according to (1.38) and (1.39) is a function of μ

and pressure P, i.e. is controlled by internal friction. The ratio $\Delta\tau_p/\tau_s$ in an original and a similar model should obviously be the same

$$\Delta\tau_p/\tau_s = \text{const.} \tag{1.41}$$

This condition is automatically fulfilled if relations (1.32) and (1.40) are satisfied. In practice, however, one has to use materials whose residual (after failure) strength τ_s^{min} is not associated with internal friction but is governed by the tixotropic properties of a material. The condition (1.41) is thus independent when using these materials, and therefore should be satisfied separately.

1.5. Conclusions

Similarity criteria (their complexity and number) are determined by the adopted theoretical models of a phenomenon under study, in the first place by models that describe the properties of the prototype (tectonic media).

1. Simple models which describe viscous (Newtonian), plastic (Von Mises), and elastic (Hooke) properties of tectonic media yield the following similarity criteria, respectively:

$$\frac{\eta^o V^o}{\rho^o g^o H^{o2}} = \frac{\eta^m V^m}{\rho^m g^m H^{m2}}, \quad \left(\frac{\eta V}{\rho g H^2} = \text{const}\right); \tag{1.42}$$

$$\frac{\tau_s^o}{\rho^o g^o H^o} = \frac{\tau_s^m}{\rho^m g^m H^m}, \quad \left(\frac{\tau_s}{\rho g H} = \text{const}\right); \tag{1.43}$$

$$\frac{E^o}{\rho^o g^o H^o} = \frac{E^m}{\rho^m g^m H^m}, \quad \nu^o = \nu^m, \quad \left(\frac{E}{\rho g H} = \text{const}; \ \nu = \text{const}\right), \tag{1.44}$$

where η, V, ρ, H, τ_s, E, and ν are the dynamic viscosity, the velocity, the density, the spatial scale, the yield limit (strength), the modulus of elasticity, and Poisson's ratio; the superscripts "o" and "m" denote original and model; $g^o = g$ where g is the acceleration of gravity; g^m is also equal to g under ordinary conditions and to the centripetal acceleration when centrifuged models are used. Criteria (1.42) to (1.44) should be used together when all properties viscous, plastic and elastic are significant for investigating process. If a property is regarded as insignificant, the relevant criterion can be dropped.

Within the framework of a linearly viscous model, time in the model is converted into "geologic" time as follows

$$t^{o} = \frac{\rho^{m} H^{m} \eta^{o} g^{m}}{\rho^{o} H^{o} \eta^{m} g^{o}} \times t^{m}.$$

The conversion is however rather conditional since a linear model is quite a rough approximation of the viscous properties of tectonic media.

2. Thermal processes that involve phase transitions have the similarity criteria

$$\frac{H^2}{\kappa t} = \text{const}; \qquad \frac{L}{C_p(T_l - T_s)} = \text{const},$$

where L is latent heat; C_p is specific heat; T_l, T_s are liquidus and solidus temperatures, respectively. If these processes develop upon the background of deformation, the appropriate mechanical similarity criteria should be added.

3. The dependence of stress τ on strain rate $\dot{\varepsilon}$ for rocks under steady state creep is

$$\dot{\varepsilon} = A\tau^{n}$$

where A is a dimensional coefficient, n = constant (n > 1). Adopting this law, which is a better approximation to tectonic and some model materials, (1.42) should be replaced by

$$\frac{1}{\rho g H} \times \left(\frac{V}{AH}\right)^{1/n} = \text{const}; \quad n = \text{const}.$$

Time is converted in this case as follows

$$t^{o} = \frac{A^{m}}{A^{o}} \left(\frac{\rho^{m} g^{m} H^{m}}{\rho^{o} g^{o} H^{o}}\right)^{n} \times t^{m}.$$

4. Incorporation of dilatancy, internal friction and hardening into the model for inelastic behaviour of the material adds three more similarity criteria to (1.43)

$$\beta = \text{const}; \;\; \mu = \text{const}; \;\; h/\tau_s = \text{const}.$$

where β is the dilatancy factor; μ is the internal friction coefficient; h is the plastic hardening modulus.

5. Brittleness of the materials is controlled by the following criteria

$$\Delta\tau_p/\tau_s = \text{const}; \;\; \Delta\varepsilon_p = \text{const},$$

where $\Delta\tau_p$ is the stress drop during failure; $\Delta\varepsilon_p$ is inelastic strain before failure.

Propagation of cracks in a brittle material is controlled by the criterion

$$K_c/(\tau_s H^{1/2}) = \text{constant}.$$

where K_c is the critical coefficient of stress intensity.

6. For rapid processes where inertial forces are large enough (are comparable with the other forces), the similarity criteria presented above should be used in conjunction with the Reynolds number $\rho VH/\eta$ = constant or, when viscosity is not controlling parameter, with one of the conditions

$$gh/V^2 = \text{const}; \qquad gt^2/H = \text{const}.$$

7. The parameters entering the similarity criteria are not necessarily constants. They can depend on different conditions. For example viscosity of the convecting linearly viscous fluid can depend on temperature: $\eta^o = \eta^o(T^o)$. This function is converted into the model using Rayleigh number as

$$\eta^m(T^m) = \frac{\rho^m \beta_v{}^m \Delta T^m g^m H^{m3} \kappa^o}{\rho^o \beta_v{}^o \Delta T^o g^o H^{o3} \kappa^m} \times \eta^o(T^o),$$

if the upper boundary temperature is assumed to be zero for both model and prototype. Appropriate model material with temperature-dependent viscosity should be used in this modeling.

8. In addition to the similarity criteria one should also meet geometrical similarity and carry out a proper transformation of boundary and initial conditions. The conditions are transformed as follows (see (1.6)): The values of the dimensionless parameters entering initial and boundary conditions remain the same, while the values of dimensional parameters are found from the condition that the ratio of such parameters to combination of the governing parameters with the same dimension, should be in the model the same as in the original.

CHAPTER 2

MODEL SETTING, TECHNIQUE AND FACILITIES

2.1. Introduction

As pointed out in the previous chapter, the main difficulties in the physical modeling are associated with the creation of model materials and experimental conditions that satisfy the physical similarity conditions. Some subduction-related processes have been already modeled in laboratory using different techniques which have allowed one to achieve different degrees of similarity between the original (prototype) and the model. *Turner* [1973] simulated subduction qualitatively. Glycerine was used to model the upper mantle. The subduction of the upper more rigid film, the lithosphere was driven by ascending gas bubbles. *Jacoby* [1976] and *Jacoby and Schmeling* [1981] simulated the upper mantle with molten paraffin. The cooled and crystallized upper layer was the lithosphere. Subduction developed due to thermal convection in the melt and gravitational sinking of the crystallized layer. *Kincaid and Olson* [1987] modeled the lithosphere and the underlying mantle using sugar syrup of various concentrations. They studied the interaction between a slab descending under its own weight and mantle transition zone at 670 km depth. The materials used allowed model parameters to be varied within sufficiently wide ranges, to make quantitative estimates possible.

In this work we applied a novel technique for the modeling of subduction developed by the author [*Shemenda*, 1979, 1981, 1984, 1985] on the basis of hydrocarbon model materials which possess properties strongly dependent on temperature. The technique, experimental facilities as well as utilized modeling approach are described in this chapter.

2.2. General Modeling Scheme

The choice of the optimal general scheme (model) for experiments is the first task to be solved. The scheme should be as simple as possible and should not at the same time omit those characteristics of the object to be studied which control the main patterns of its evolution. Once a scheme has been determined, the modeling should strictly obey similarity conditions. There is little sense, for instance, in studying complex inhomogeneous, multi-layer models that appear to be similar to the natural object, but which do not satisfy similarity criteria. There is no any guarantee that the results of such modeling would correspond to the actual process to any degree. On the other hand, an oversimplified model that does not involve some essential features of structure and dynamics of the prototype will also not yield new knowledge on the object, even though the similarity criteria have been followed strictly.

The model of subduction must be at least two-layered including strong (very viscous) lithosphere and underlying effectively liquid asthenosphere. It is more or less clear that the asthenosphere should be modeled by a low viscosity liquid, whereas it is not obvious at all which rheologic model can be assumed to describe the effective behaviour of the lithosphere in subduction zone. The actual properties of the oceanic lithosphere at deformation rates typical of a subduction zone are controlled largely by the following factors: (1) a brittle behaviour of the upper plate layers which have relatively low strength; (2) a gradual transition with depth from brittle to brittle-dilatant and to plastic properties with a considerable increase in effective strength (yield limit) σ_S; (3) a decrease in σ_S when a depth of a few tens of kilometers has been reached (depending on lithospheric age), and the transition to high-temperature creep at larger depth [*Ashby and Verrall*, 1978; *Goetze and Evans*, 1979; *Kirby*, 1980, 1983; *Nikolaevsky*, 1983]. Such rheologic stratification could be incorporated into a three-layered lithospheric model in which the upper weak layer is brittle-dilatant (semi brittle), the intermediate strongest layer is plastic, and the lower layer is viscous. If one is interested only in large scale deformation of the lithosphere then it is not necessary to pay much attention to the details of its rheologic stratification. The plate can be considered as one-layered. Such a lithospheric model should probably have plastic or elastico-plastic properties since the main, thickest and strongest central (core) layer of the actual lithosphere is plastic. The quantitative parameters for this lithospheric model, the yield limit σ_s and the modulus of elasticity E, must obviously differ from those derived from the experiments with rocks. Such experiments do not take into account spatial and time scale for the geological processes. The parameters must have some effective values which can be derived indirectly (using modeling) from some geological and geophysical data (in particular, gravity anomalies) which characterize the response of the lithosphere to loading on both geological time and spatial scales.

One-layer elastico-plastic lithospheric model (with series connection of plastic and elastic elements, see Section 1.3.5) seems to provide a good approximation for studying the subduction of oceanic lithosphere and is used in Chapters 3, 4 and 5. However, this model is probably insufficient for modeling the subduction of the continental lithosphere in the collision zones (continental subduction). Due to the large thickness of the continental crust its lower layer may be very weak [*Ranalli and Murphy*, 1987] resembling mantle asthenosphere. The upper crust and the mantle part of the continental plate separated by this weak layer, the "crustal asthenosphere" [*Turcotte et al.*, 1984], can deform in different ways. Therefore for modeling the continental subduction presented in Chapter 6 we use three-layer lithospheric model which includes a semiplastic-semibrittle upper crustal layer, a ductile, very weak crustal asthenosphere, and a plastic mantle layer.

The choice of a model for the asthenosphere is defined by the fact that its effective strength σ_s or viscosity η is several orders less than that for the lithosphere. The low σ_s and η define the low tangential stresses τ_a (friction) which the asthenosphere exerts on the lithospheric base. These stresses are estimated at 10^6 Pa [*Artushkov*, 1979], while the stresses within the lithosphere in subduction zones reach its effective yield limit σ_s which is estimated of the order of 10^8 Pa [*Lobkovsky and Sorokhtin*, 1976; *Ushakov*, 1968]. Therefore, the stress τ_a can be neglected when studying the lithospheric deformation within the area of a subduction zone. Accordingly, the yield limit or viscosity coefficient for the asthenosphere can be set to zero. The role played by the asthenosphere in such an approximation is only to make Pascal's law hold beneath the lithosphere, i.e., to maintain hydrostatic equilibrium under the lithosphere. This approximation of course is fairly rough and completely disregards a viscous interaction between the subducted plate and the surrounding mantle which actually increases with depth. The viscosity of the asthenosphere is related also to the problem of a driving mechanism for subduction. The main sources for plate motion are believed to be the sliding of the lithosphere from mantle rises (ridge-push), traction of the lithosphere produced by mantle (asthenospheric) currents, and gravitational sinking of the lithosphere into the mantle in the subduction zones (slab-pull). In the first mechanism (ridge-push) a force pushing the slab into the subduction zone originates at the mid ocean ridge due to lithospheric curvature, and is transmitted along the lithosphere to the subduction zone. In the second mechanism (mantle traction), slab compression in the subduction zone is generated by stresses τ_a acting on the lithospheric base. The stresses are small but being summed over a vast area can produce large compressive stresses at some part of the lithosphere where a subduction zone can develop. Directly in the subduction zone, stresses τ_a do not play an important role, being very small. Therefore, the boundary condition $\tau_a = 0$ is a good approximation for the lithosphere deforming in a subduction zone. It is only the net effect of τ_a over a large area outside the subduction zone which is important.

The first two mechanisms can readily be replaced by local application of lateral compression to the slab using a rigid piston. The compression should be applied far enough from the subduction zone (or place of its formation) for the boundary effects related to the loading method to be negligible. The third driving mechanism, the slab-pull can be modeled directly by an increase in the density of the model lithosphere.

Thus, for modeling the subduction of the oceanic lithosphere we assume simple two-

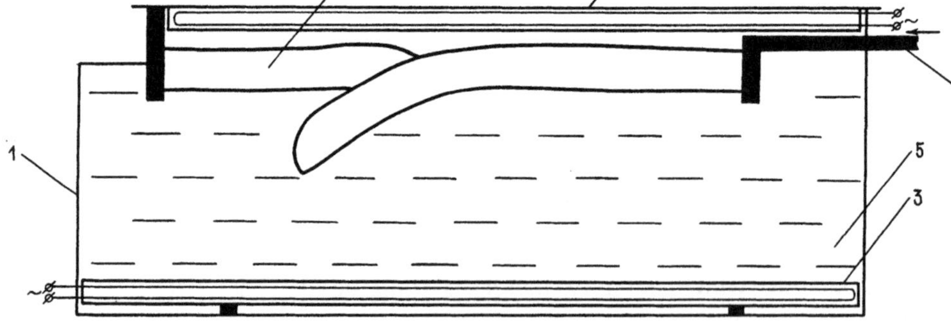

Figure 2.1. Scheme of the model experiments and the experimental installation 1, tank; 2, "lithosphere"; 3, electric heaters; 4, piston; 5, "asthenosphere".

layer model (Figure 2.1) which includes elastico-plastic lithosphere of variable density and a very low-viscosity asthenosphere. The lithosphere is subjected to horizontal compression by means of a piston. For modeling continental subduction we use a three-layer model for continental lithosphere and the same very low-viscosity asthenosphere (see Chapter 6).

Similarity criteria for subduction of the oceanic lithosphere within the described scheme follow directly from (1.24) (1.43), and (1.44)

$$\tau_s/(\rho_l gH)=\text{const};\ E/(\rho_l gH)=\text{const};\ \rho_l/\rho_a=\text{const};\ Vt/H=\text{const} \qquad (2.1)$$

where τ_s, E, H, and ρ_l are the shear yield limit, the modulus of elasticity, thickness and density of the lithosphere, respectively; ρ_a is the asthenospheric density; V is the slab (piston) velocity; t is the time; and g is the acceleration of gravity. Neglected here are Poisson's ratio of the lithosphere and the pressure of ocean water on the plate surface. It should be noted that the last condition in (2.1) has a purely kinematic meaning and serves only to convert model time into geologic time (see Section 1.3.1). The condition imposes no restriction on the process rate in the model (see Section 1.3.3).

2.3. Model Materials

Substitution of the parameter values typical of the prototype and realistic (for the laboratory conditions) values of model plate thickness and density into (2.1) shows that the yield limit for the model lithosphere must be very small, of the order of 10 Pa. A material having such strength flows under its own weight and breaks when held in the hand. This is one of the main technical problems in the modeling large-scale tectonic processes which makes it difficult to prepare specimens of model lithosphere of required shape and to perform mechanical operations with them before and after the experiment. This problem can be resolved by utilization of special model materials which change their properties according to change of confining conditions. In this work we use a specially developed systems of composite materials consisting of alloys of solid hydrocarbons (paraffins and ceresins), mineral oils, finely ground powders, and a small addition of surface-active substances [*Shemenda*, 1981, 1984, 1989*c*]. Structurally, these systems are thixotropic dispersions of solid hydrocarbons and powders in oil. They posses complex elastico-viscoplastic properties (see Figures 2.2 and 2.3) that are strongly dependent on the composition, temperature and conditions of preparation of the models (conditions of

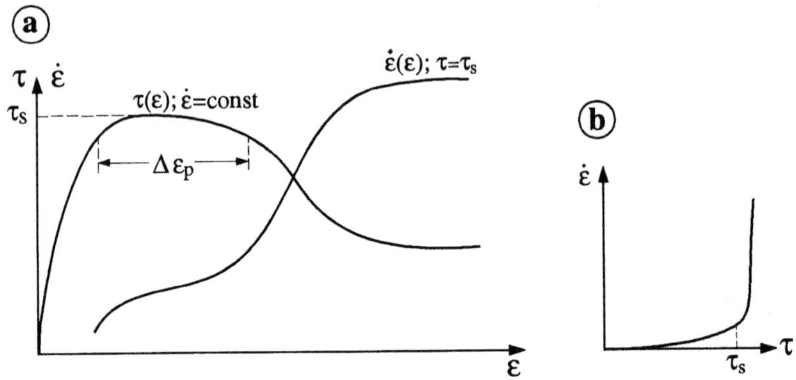

Figure 2.2. Properties of hydrocarbon model materials (shear testing).
(*a*) Summary relationships $\tau(\varepsilon)$ for $\dot{\varepsilon}$ = const and $\dot{\varepsilon}(\varepsilon)$ for $\tau = \tau_S$;
(*b*) Summary relationship $\dot{\varepsilon}(\tau)$;

structuring for the material), and are virtually independent of confining pressure. One can vary the composition and conditions to obtain qualitative and quantitative properties of the model lithosphere (as well as crustal and sedimentary layers) that would satisfy the similarity criteria.

The micromechanism for deformation of hydrocarbon compounds is associated with failure in the structural frame of the material and its molecular recovery. Owing to this mechanism, the model materials exhibit different deformational properties for different ranges of temperature T and strain rate $\dot{\varepsilon}$, from linear viscous to nonlinear viscous, to plastic and brittle-dilatant properties. For the modeling presented in this work we choose ranges for T and $\dot{\varepsilon}$ such that the yield limit τ_s on the one hand satisfies similarity criteria (2.1) and on the other hand is weakly dependent on $\dot{\varepsilon}$, which means that the material corresponds to the accepted model of a elastico-plastic solid. The composition of the material was chosen such that the temperature at which the properties are elastico-plastic and satisfy similarity criteria would not be excessively high and that at the room temperature the material be strong enough to permit mechanical operations with the models without a risk of damaging them. All these conditions have been satisfied by using the following compounds for modeling the lithosphere, the crust, and the sedimentary layer, respectively:

Compound N1: paraffins - 7.1%, ceresins - 8.8%, mineral oils - 61.7%, finely ground powders - 23-30% (for modeling the oceanic lithosphere and mantle layer of the continental lithosphere).

Compound N2: paraffins - 9.2%, ceresins - 11.4%, mineral oils -79.4% (for modeling the crust).

Compound N3: paraffins - 4.2%, ceresins - 9.3%, mineral oils - 86.5% and

Compound N4: paraffins - 3.3%, ceresins - 6.5%, mineral oils - 91.2% (for modeling the crustal asthenosphere and sediments).

The properties of the compounds are represented by curves in Figure 2.3. The $\tau(\varepsilon)$ curves in Figure 2.3*b*, for example, were obtained at *T* and in the range of $\dot{\varepsilon}$ used in the model experiments. The curves demonstrate really a minor dependence of τ_s on $\dot{\varepsilon}$ in the plastic flow regime. At the same time, Figure 2.3*b* reveals a rather strong dependence of the inclination of the subvertical segments of the τ(s) curves on $\dot{\varepsilon}$. This permits one to vary $\dot{\varepsilon}$ in order to modify the effective modulus of elasticity *E* of the model material (*E* is controlled by the slope of the subvertical segments). The model *E* is of the order of 10^3 Pa (Figure 2.3*b*) and when converted to nature is equal to 10^{10} Pa.

For modeling the asthenosphere we used different low-viscosity fluids, but the most convenient and suitable material has proved to be pure water. Its density is 10^3 kg/m^3. The density of hydrocarbon systems for the model lithosphere can be made either lower, higher or equal to this value by variation of concentration of the powder.

2.4. Experimental Installation and Technique

A strong dependence of τ_s on *T* for the model material prescribes the construction of the experimental installation and the modeling technique. A diagram of the experimental apparatus is shown in Figure 2.1. An important component of the installation is a thermostatic system maintaining a specimen of model lithosphere at a constant working temperature (temperature of experiment at which properties of the model material corresponds to the similarity conditions). To perform an experiment, water simulating the asthenosphere is poured into a tank (1, Figure 2.1) with sizes 40 × 8 × 20 cm for 2-D modeling and 40 × 40 × 20 cm for 3-D experiments. A model lithospheric plate is placed

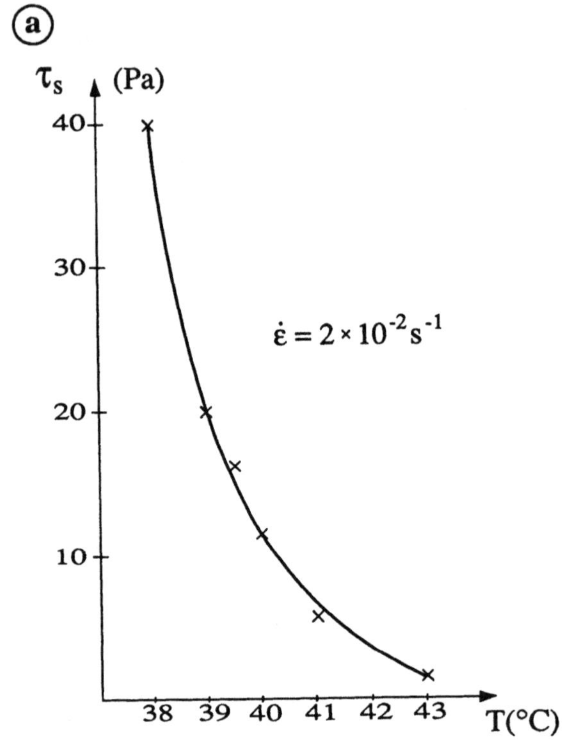

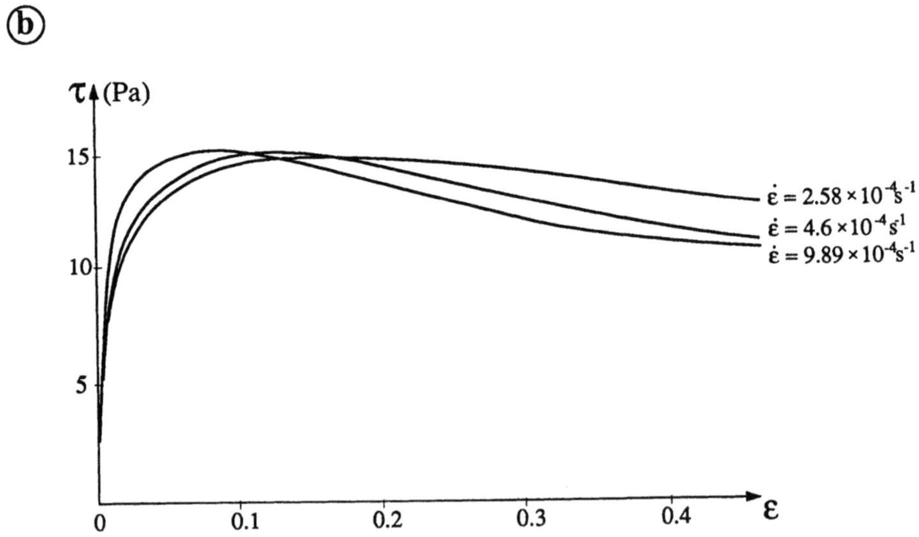

(see figure caption on the next page)

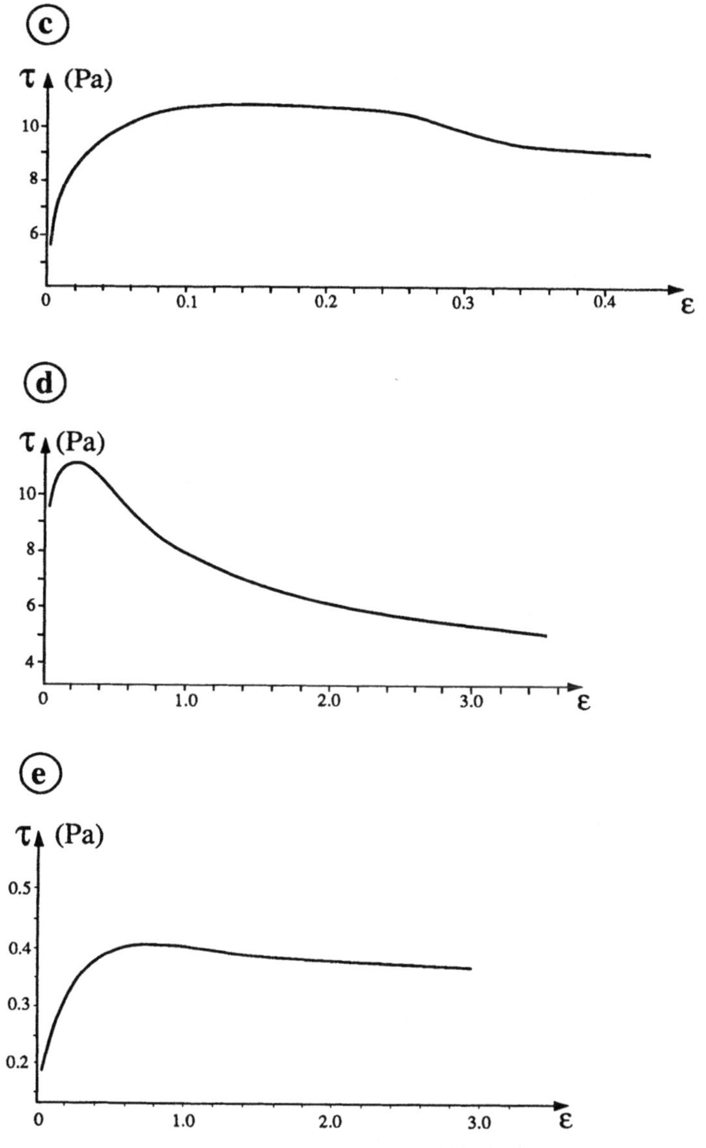

Figure 2.3. Curves characterizing properties of the model materials (shear testing)

(*a*) $\tau_s(T)$ for compound N1 at $\dot{\varepsilon} = 2 \times 10^{-2}$ s^{-1};

(*b*) $\tau(\varepsilon)$ curves for compound N1 at $T = 39.5°C$ and different $\dot{\varepsilon}$;

(*c*) and (*d*) $\tau(\varepsilon)$ curves for compound N1 at $T = 40°C$ and $\dot{\varepsilon}$ = const; (c) $\dot{\varepsilon} = 2.5 \times 10^{-3}$ s^{-1}; $\varepsilon < 0.5$; (d) $\dot{\varepsilon} = 2 \times 10^{-2}$ s^{-1}, $\varepsilon < 4$;

(*e*) $\tau(\varepsilon)$ curve for compound N3 at $T = 40°C$ and $\dot{\varepsilon} = 2 \times 10^{-2}$ s^{-1}.

above. The plate specimen is in direct contact with the tank walls, but special measures are taken to reduce the friction between the plate and the bath walls to nearly zero. Plane electric heaters (3) are used to make the entire system thermostatic at the working temperature 38° to 42°C (39.5° for most experiments described below). The thermostating process lasts about 2 hours and is monitored using thermocouples which can be inserted into the plate itself. Piston (4) is then used to produce compression of the lithosphere. Plate deformation is studied through the transparent bath walls and by using the following technique for visualizing the deformation: The specimen is frozen before the experiment and then is cut longitudinally at the middle into two parts. The section of one part is treated with a special stamp to impress a grid matrix with lines 0.1 mm wide at intervals of 2 mm. The matrix is filled up with a paint suspension. The two halves are then re-attached together. The specimen is placed into the installation, heated, and deformed. The experiment can be stopped at any required stage and the model is allowed to cool. The solidified deformed specimen is then removed from the bath and cut again along the previous section. The displayed grid allows us to derive the finite deformation and to identify faults. Circles 3 to 5 mm in diameter are also stamped into the slab. The deformation of the circles are used to derive the directions of the principal strain and stress axes.

We applied the following experimental procedure to determine the normal horizontal non-(super-)hydrostatic stress σ_h in the "lithosphere" during subduction (unfortunately these stresses cannot be measured instrumentally as they are very small). A series of experiments was conducted under the same conditions. Prior to the experiments, the model lithosphere was thinned at a certain location from below (a notch with width comparable with the specimen thickness H was made in the slab base all the way across it perpendicular to the direction of loading). The height of the notch was increased gradually from one experiment to the next, until failure occurred at the weakened location. Knowing the maximum thickness h_m of the thinned location where failure occurred, we determined σ_h, $\sigma_h = \sigma_s \times h_m/H$, where σ_s is the yield limit at normal load ($\sigma_s \approx 2\tau_s$ for plane strain state [*Kachanov*, 1969] which takes place in our case).

CHAPTER 3

INITIATION AND DEVELOPMENT OF SUBDUCTION UNDER HORIZONTAL COMPRESSION OF THE LITHOSPHERE

3.1. Introduction

More then 80% of seismic energy in subduction zones is released in thrust type earthquakes with the axis of maximum compression oriented transverse to the subduction zones [*Balakina*, 1984]. This is consistent with the convergence of the lithosphere in these zones. In spite of this, it is not obvious that the general stress regime of the lithosphere in a subduction zone is characterized by horizontal compression. It is well known that the lithosphere in subduction zone can undergo tension and moreover, the convergence itself can be driven by the gravitational sinking of the subducting plate into the mantle. However, to form the subduction zone, to start the convergence and thinking of the plate into the mantle, large compression of the lithosphere is probably necessary. It seems that there is no other physically realistic way to cause failure of the plate and to initiate subduction other than application of large compressive stresses to the lithosphere. This chapter is devoted to the investigation of possible mechanisms for formation of subduction zones and some general patterns of the subduction process.

3.2. Results of the Experiments

Several hundred experiments have been carried out under different conditions (geometry and composition of the model lithosphere, working temperature, and strain rate) [*Shemenda*, 1979, 1989*a*, and 1992]. Experiments were repeated several times under the same conditions to exclude random results. Below we describe some of the most typical experiments for which densities of the lithosphere and asthenosphere are equal. Unless otherwise stated, these experiments were conducted with the following model parameters: $\tau_s^m \approx 15$ Pa; $E^m \approx 2 \times 10^3$ Pa; $H^m = 2.3 \times 10^{-2}$ m; $\rho_l^m \approx \rho_a^m = 10^3$ kg/m^3 which satisfy similarity criteria (2.1) if parameters for the prototype are the following: $\tau_s^o = 1.3 \times 10^8$ Pa; $E^o = 1.7 \times 10^{10}$ Pa; $H^o = 6 \times 10^4$ m; $\rho_l^o \approx \rho_a^o = 3.3 \times 10^3$ kg/m^3 (here τ_s, E, H, and ρ_l are the shear yield limit, the modulus of elasticity, thickness and density of the lithosphere, respectively; ρ_a is the asthenospheric density; V is the slab (piston) rate; t is the time; and g is the acceleration of gravity). The above listed values for the prototype are within the available estimates for the parameters of the lithosphere and asthenosphere that can be considered as a formal confirmation of the adopted model parameters. As a matter of fact, these values (for the prototype) were determined in the course of modeling itself. This will become clear toward the end of this chapter where we discuss the experimental results and actual situations. The average subduction rate in the model is $V^m = 2 \times 10^{-5}$ m/s. Taking a mean value of $V^o = 5$ cm/yr for the subduction rate in nature, we can use the last condition in (2.1) to derive that 1 min in the model corresponds to about 6×10^4 years in nature.

Experiment 3.1. The model of a homogeneous oceanic plate is subjected to horizontal compression (Figure 3.1). At the initial stage of the experiment, the model experiences more or less uniform plastic shortening along its entire length. Numerous shallow reverse faults perpendicular to the axis of compression appear on the initially smooth plate surface. Major thrust fault zones crossing the whole plate thickness then form near the piston and the opposite tank wall (Figure 3.1*b*). Simultaneously or somewhat later, the specimen loses buckling stability and bends. (The bending and generally, the vertical displacements of the model in this and in the all following experiments can be observed with respect to the horizontal black line drawn on the bath wall slightly below the plate surface. The surface is parallel to the line before deformation). Further compression produces "vertical" subduction along one of the thrust fault zone, usually near the piston (Figures 3.1*c* to 3.1*f*). The subduction develops through the successive breaking out of model blocks along inclined shear (thrust) faults and "trickling down" of these blocks along the piston.

Experiment 3.2. The lithospheric model is thinned locally before experiment (Figure 3.2). Deformation of such a model is localized in the weakened zone at the onset of compression. At first, the "lithosphere" in this zone undergoes some contraction accompanied by shallow thrusting, similar to the initial stage of experiment 3.1. The thinned segment then rises (Figure 3.2*b*). Thrusting at the margins of the rise becomes particularly intensive. Then, two major, plastic, thrust zones form at these margins. They

Figure 3.1. Result of experiment 3.1. Horizontal compression of the oceanic lithosphere model.

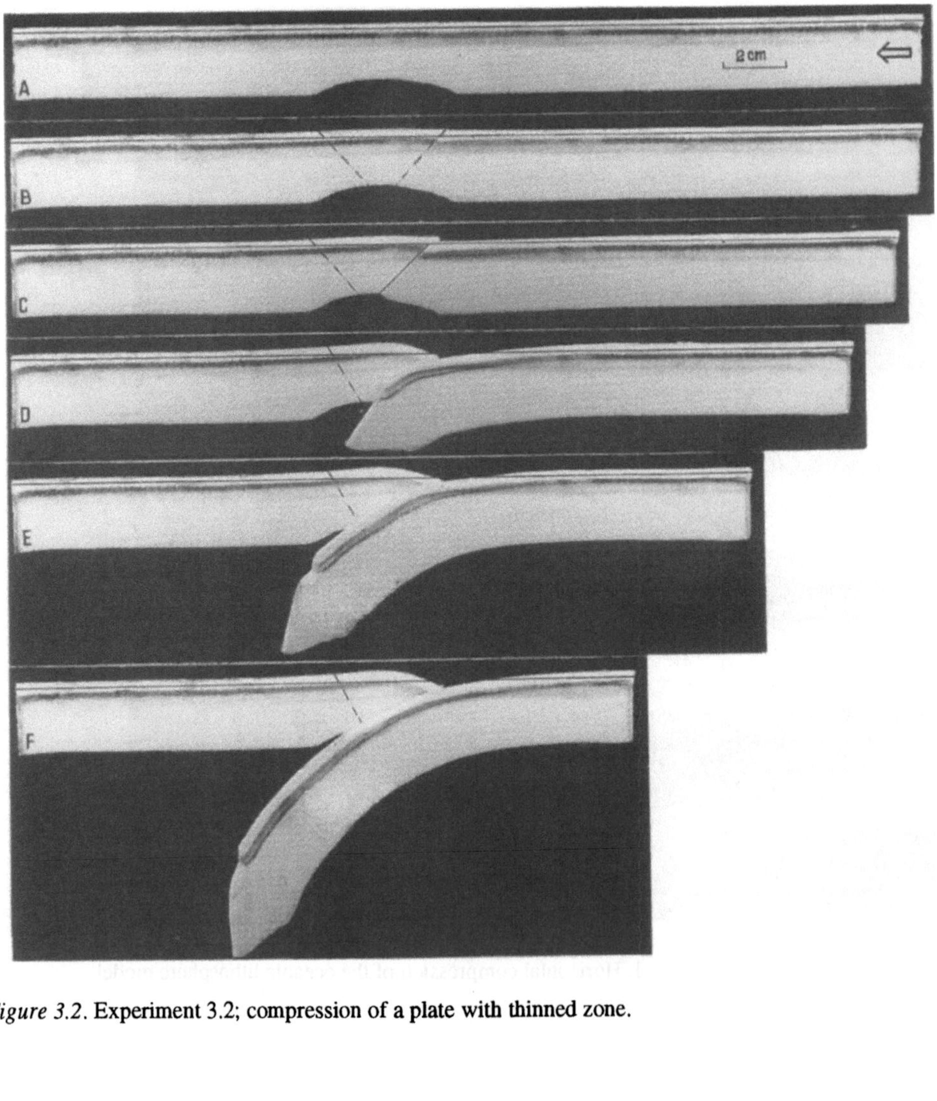

Figure 3.2. Experiment 3.2; compression of a plate with thinned zone.

cross the whole plate thickness and dip toward the center of the weakened segment. Complete failure of the lithosphere can then occur along one of these zones with equal probability. Further compression produces subduction of one fragment of the broken slab under the other (Figures 3.2*d* to 3.2*f*). The dip angle of the thrust zone is close to 45° (in cross section) and so is the angle of initial subduction. The angle of subduction, however, decreases during subduction (Figures 3.2*c* to 3.2*f*) due to nonuniform scraping (erosion) of the material from the base of overriding wedge. The subducting lithosphere is bent down in front of subduction zone producing a trench (Figures 3.2*d* to 3.2*f*). An extended rise of very small amplitude forms to the right of it. This is the "outer rise" or flexural bulge which is seen more clearly in the next experiments. Narrower but higher rise also forms to the left of the trench. This rise can be identified as the frontal (nonvolcanic or nonisostatic) arc.

Experiment 3.3. The model lithosphere consists of two segments with different thicknesses (the material is the same), Figure 3.3. The thicker segment can be treated as continental lithosphere and the thinner one as oceanic lithosphere. Thus the whole model can be considered as the lithosphere in the transitional area between continent and ocean. Compression of this model first results in uniform shortening of the thin (oceanic) plate and thrusting in its upper part similar to experiment 3.1. The oceanic plate then loses stability and buckles (Figure 3.3*b*). A swell forms in the zone of thickness contrast (i.e. directly in the "continental margin") which is followed by downwarp farther to the "ocean" (to the left), Figures 3.3*b* and 3.3*c*. The amplitude of flexural buckling w then increases while the wavelength λ decreases. At the next stage whole contractional deformation concentrates within the downwarp where the lithosphere undergoes a violent plastic compression and thrusting (Figures 3.3*c* and 3.3*f*). Both a fractured ridge at the plate surface and a more massive "lithospheric" root develop in the zone of deformation localization (Figures 3.3*f* and 3.4*b*). During this process two major intersecting plastic thrust faults develop through the whole plate thickness (Figures 3.3*f* and 3.4*b*). Further compression results in complete failure of the lithosphere along one of these faults and initiation of subduction (Figures 3.3*d*, 3.3*e*, 3.3*g*, and 3.4*c*). We measured the normal nonhydrostatic stresses σ_h in the lithosphere in experiment 3.3 using the technique described in Section 2.4. The results have shown that at the stage of flexural buckling of the oceanic plate, σ_h is close to the lithospheric yield limit for normal loading σ_s. Further deformation (during the stage of deformation localization) results in gradual failure of the material, causing σ_h to decrease and approach a value of about (0.3-0.4)σ_s which is the

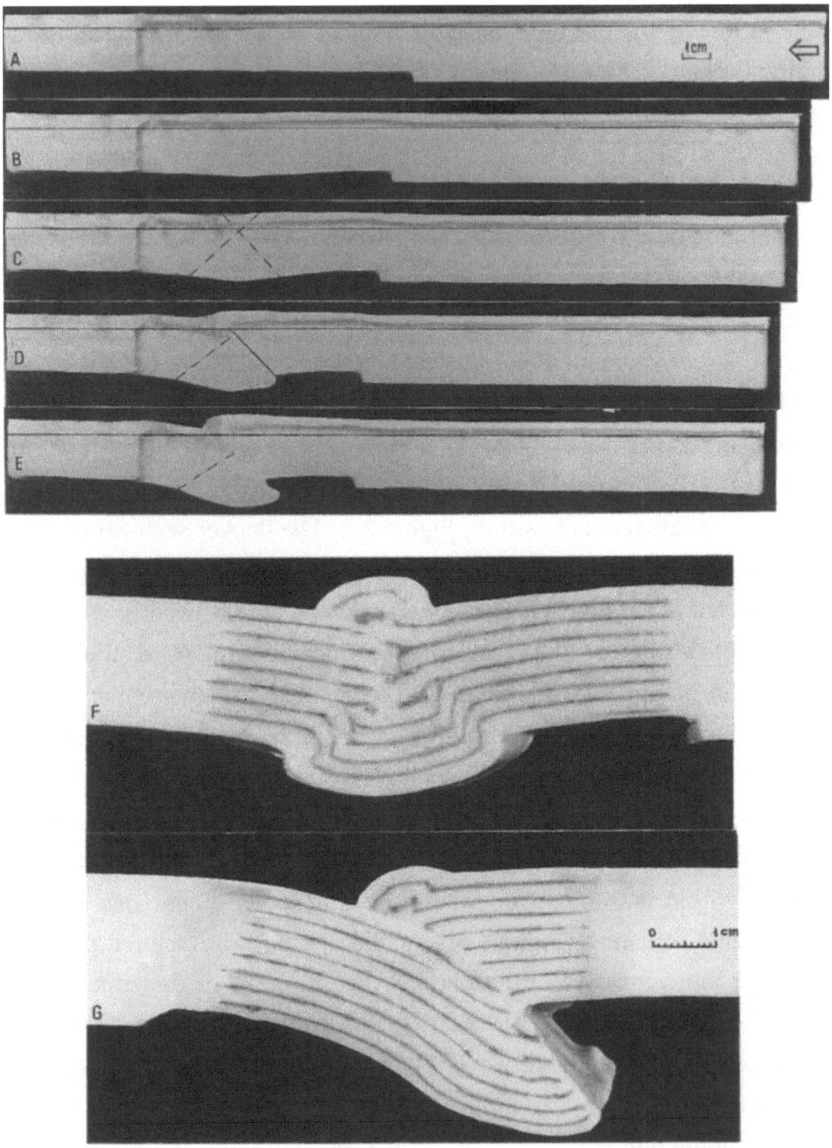

Figure 3.3. Experiment 3.3; compression of the model of a transition zone from ocean to continent (the continuous lithosphere model consists of two parts of different thicknesses)
(*a*) to (*e*) Stages of experiment (general view through the wall of the experimental installation); (*f*) and (*g*) Cross sections of the frozen models corresponding to the stages of (*f*) deformation localization and (*g*) plate failure and initiation of subduction.

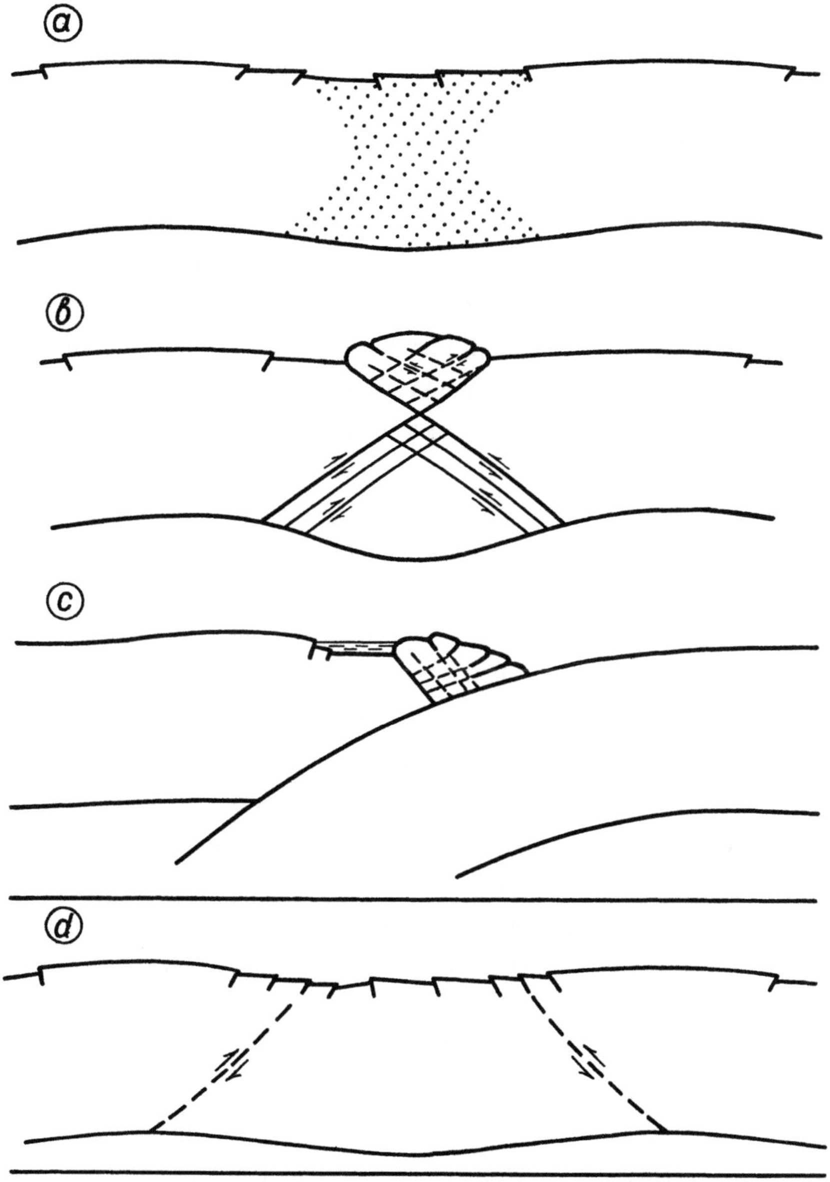

Figure 3.4. Scheme of deformation localization (from experiments 3.3, 3.4, and 3.5)
(*a*) to (*c*) Stages of deformation localization; (*d*) Another possible variant for the deformation localization and failure of the lithosphere (this case corresponds to the reduced thickness *H* and increased τ_s); stippled in Figure 3.4*a* is zone of violent plastic compression (shortening).

stress necessary to maintain the subduction.

We also conducted similar experiments with other parameters of the model and found that the process of deformation localization depends on several factors, in particular on the yield limit τ_s of the model material and its brittleness (on plastic deformation $\Delta\varepsilon_p$ prior to failure, see Figure 2.2*a*). An increase in $\Delta\varepsilon_p$ leads to more intensive plastic "squashing" of the plate and to formation of both a broader ridge and lithospheric root. Complete failure of the lithosphere in this case occurs after greater deformation. A decrease in τ_s has roughly the same effect causing a decrease in the maximal amplitude w of plate flexure (the plate fails at smaller w). On the other hand, increasing the yield limit modifies the failure mechanism. The localization of deformation still occurs in the downwarp, but maximal concentration of deformation develops rather at the margins of the downwarp than at the center. Two major thrust (shear in the cross section) zones form at these locations, followed by complete failure of the plate along one of them (Figure 3.4*d*) and initiation of subduction. The same result can be obtained by decreasing the plate thickness with the same yield limit as in experiment 3.3.

It is of interest that the pattern of model deformation obtained in experiment 3.3 (Figure 3.3) remains practically the same, if the "continental" and "oceanic" segments of the lithosphere are separated with a vertical cut made beforehand. The only difference is that a small vertical offset occurs during the experiment between the segments along the cut. This is due to the different amplitude of uplift of these segments (the thin plate rises higher than the thick one during flexural buckling). If however, the cut is inclined then subduction develops along this cut before compressive stresses reach the value needed for buckling and initiation of plate failure in a new place. There is a critical value for the cut dip angle which if exceeded, changes mode of deformation (results in immediate subduction along the cut).

Experiment 3.4. The oceanic plate is made longer and the continental plate shorter in this experiment (Figure 3.5). The plates are separated with a vertical cut. The result is similar to experiment 3.3. Experiment 3.4 clearly displays a flexural buckling of the oceanic plate. The deformation localization develops as in experiment 3.3 in the sag closest to the continental margin.

Experiment 3.5. In this experiment we tried to take into account the weight of sedimentary material accumulating at a passive continental margin. This material exerts considerable pressure on the underlying lithosphere and may influence the process of subduction initiation [*Cloeting et al.*, 1989]. The model "sedimentary" wedge (dotted in

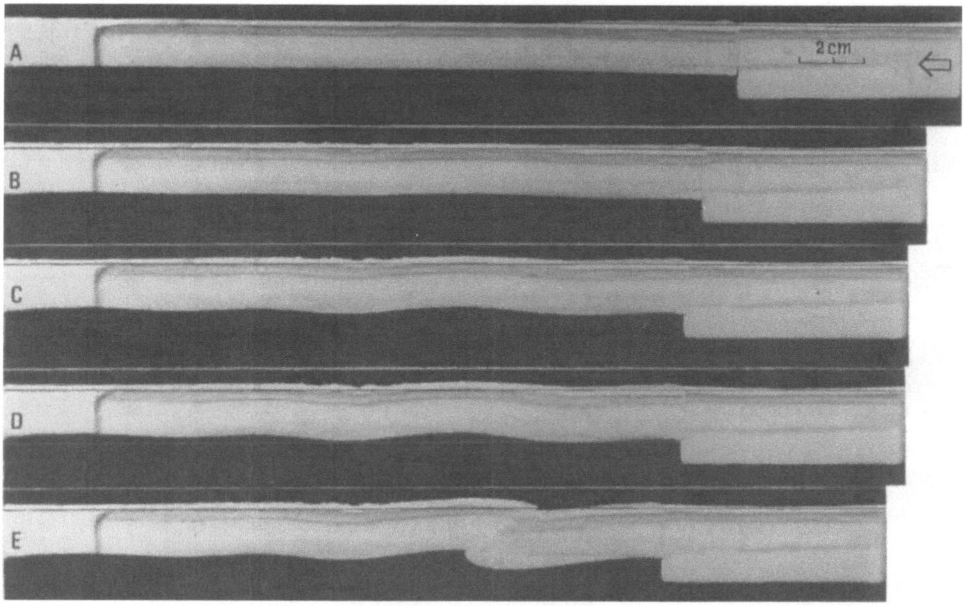

Figure 3.5. Experiment 3.4; this experiment differs from the previous by the longer length of the oceanic lithosphere; in addition, the continental and oceanic lithosphere are separated by a vertical cut.

Figure 3.6*a*) is made of compound N3 which is several times weaker than the lithosphere (see Section 2.3). The parameters for the model wedge correspond to the following parameters for the prototype: density $\rho_s^o = 2.5 \times 10^3$ kg/m^3; height of the base $h_s^o = 6$ km; length $l^o = 150$ km. The wedge is placed on the oceanic slab which is separated from the continental lithosphere by the vertical cut. The slab bends under the wedge weight. Compression of this model results in further bending (buckling) of the oceanic plate. In contrast to the preceding experiments where a swell forms directly at the transition zone from oceanic to continental lithosphere, in experiment 3.5 this area experiences further

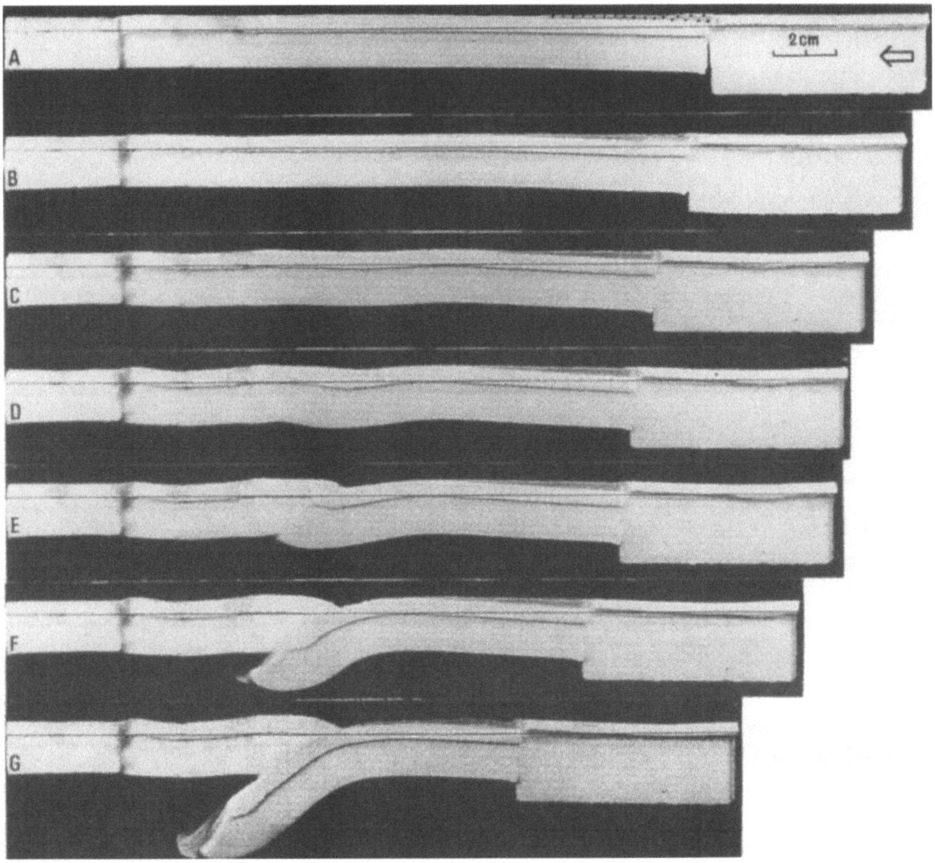

Figure 3.6. Experiment 3.5; this experiment differs from the previous by the presence of a "sedimentary" wedge (stippled in Figure 3.6*a*) on the oceanic plate in the "passive continental margin".

subsidence enforced initially by the sedimentary load. In other words, the phase of slab flexure has been shifted by π due to the effect of sediments. Accordingly, the depression in which deformation is localized and subsequent failure of the lithosphere occurs, has shifted seaward (Figures 3.6*d* and 3.6*e*). The same picture in principle persists if there is no vertical cut at the continental margin. The only difference is that both oceanic and continental plates are involved in downwarping which in this case is less deep.

There is a critical value for the thickness of the sedimentary wedge h_s below which the phase of slab flexure is not shifted. When h_s is diminished by a factor of ~ 1.5 to 2 compared with experiment 3.5, the phase shift no longer occurs. There is also some amount of "sedimentary" material at the continental margin in experiment 3.4 (Figure 3.5), but it causes almost no effect on the deformation of the lithosphere.

It should be noted that in experiments 3.3 to 3.5 compressive stress in the oceanic plate (along its entire length) at the stage of buckling is close to the yield limit σ_s. In some experiments with the conditions similar to experiments 3.3 to 3.5 the oceanic plate failed in different locations, including near the bath wall and directly in the transition zone under the "sedimentary" wedge. However, the most typical results were as described above.

In the next three experiments (experiments 3.6 to 3.8) the initially homogeneous lithosphere is cut along an inclined plane to separate the lithospheric model on subducting and overriding plates. The cut is made at a lower angle than the faults that formed in the preceding experiments. This was done to imitate later mature subduction stages for which the contact zone between the plates is typically flatter. Experiments 3.6, 3.7, and 3.8 differ by the lithospheric yield limit value.

Experiment 3.6. In this experiment the slab is relatively strong, τ_s^m = 40 Pa. Higher strength leads to larger vertical displacements of the plates in subduction zone (Figure 3.7). The trench depth in this experiment converted to nature amounts to several tens of kilometers which is unrealistic. The width of the trench is also too large.

Experiment 3.7. The model lithosphere has a low strength τ_s^m = 1.2 Pa. In this experiment the vertical displacements of the lithosphere model are too small (Figure 3.8). The trench is shallow, being equivalent to only a few kilometers (depth with respect to the surrounding seafloor) in nature. Although this value is realistic, the trench is too narrow in this experiment. Its width converted to nature is approximately 60 km which is about half of the width of real trenches. In addition, the distance from the trench to the frontal arc and the width of the outer rise are too small.

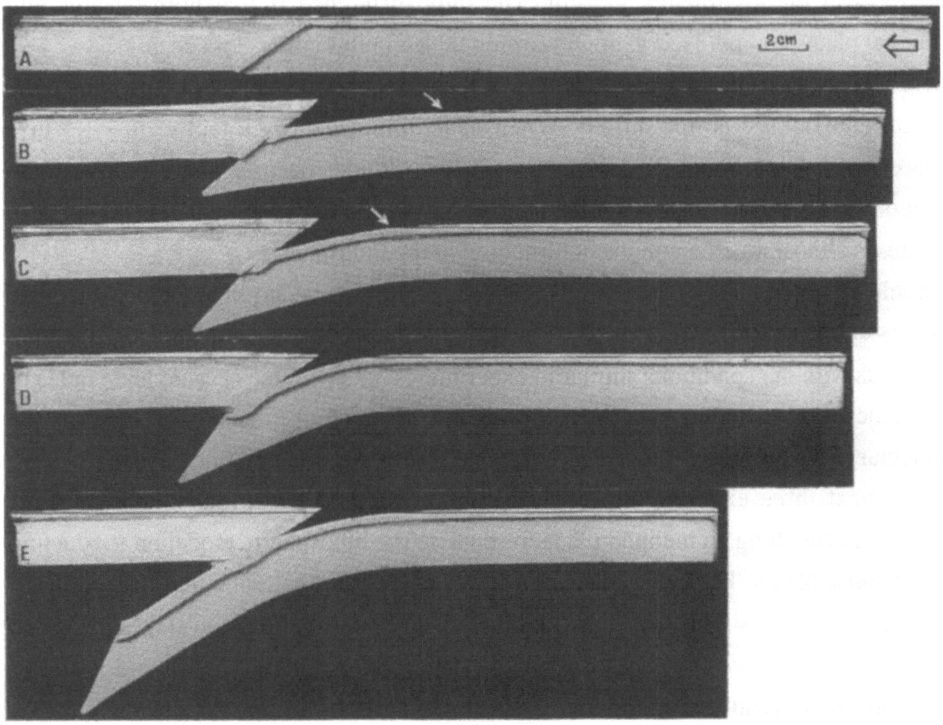

Figure 3.7. Experiment 3.6; compression of a lithospheric model with a transverse cut though the plate dipping at an angle of 35°; τ_s^m = 40 Pa which is higher than in the previous experiments. The location of the maximal bending of the subducting plate is shown in Figures 3.7*b* and 3.7*c* by white arrows (for other explanations see the text).

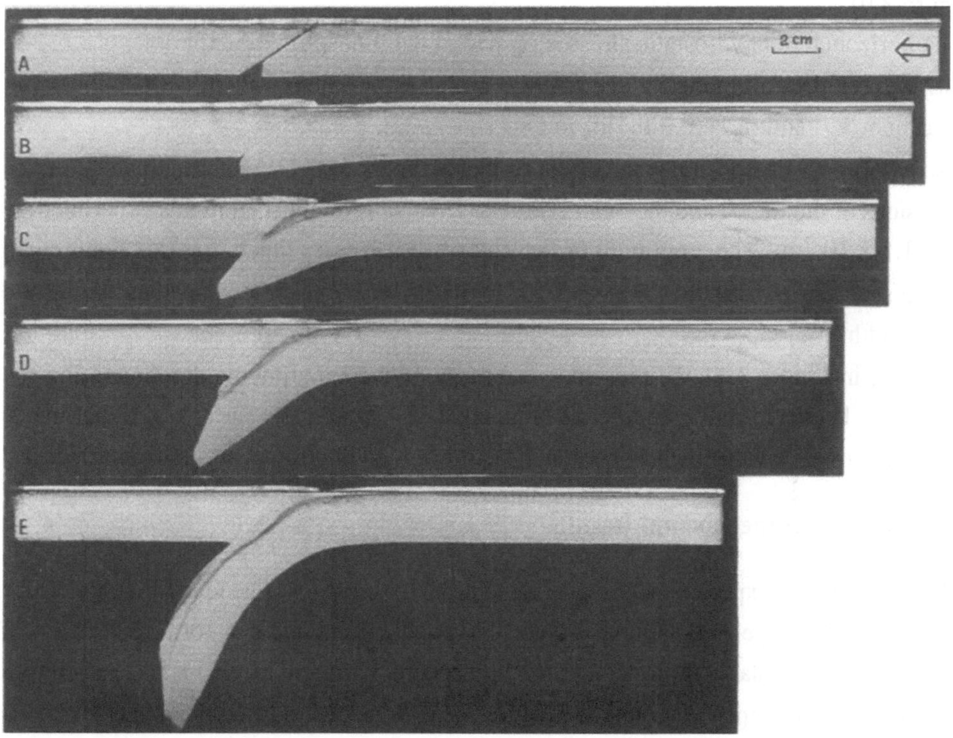

Figure 3.8. Experiment 3.7 (this experiment is characterized by a lower τ_s^m value: τ_s^m = 1.2 Pa.

Increasing τ_s^m compared to its value in experiment 3.7 improves the parameters of topography in subduction zone approaching them to the actual values. The outer rise and the trench become wider. The width of the frontal arc also increases, while its apex is displaced toward the "back arc basin". There is some increase in trench depth and possibly in the height of the outer rise (this is difficult to detect owing to its very small amplitude), the bending of the subducting plate in subduction zone becomes smoother.

Experiment 3.8. The most proper value for τ_s^m (which provides the best fit to the generalized relief in subduction zone) is about 15 Pa which was used in this experiment (Figure 3.9).

The horizontal nonhydrostatic stresses σ_h within the lithosphere in experiments 3.6, 3.7, and 3.8 have the following values: $\sigma_h \approx 0.2\sigma_s$ in experiment 3.6; $\sigma_h \approx (0.7\text{-}0.9)\sigma_s$ in experiment 3.7; and $\sigma_h \approx (0.3\text{-}0.4)\sigma_s$ in experiment 3.8.

Experiment 3.9 demonstrates the effect of the thickness of the subducting plate H on the dimensions of the trench and the outer rise. The H value is diminished in this experiment to $H^m = 1.5 \times 10^{-2}$ m. The amplitude of the outer rise increases (such that this feature can now be seen more distinctly, Figure 3.10) while its width becomes narrower as did the trench width.

Finally, in Figure 3.11 we present a summary of the experiments demonstrating the relationship between the topography and dip angle of the interplate surface ψ. Experiments shown that lower ψ value makes the trench shallower and the frontal arc lower and wider.

3.3. Analysis of Experimental Results

The results of the experiments provide some ideas for creation of geodynamic models for subduction. The experiments have shown that subduction zone can form as a result of considerable horizontal compression of the lithosphere. Different mechanisms can produce a subduction zone, depending on the structure of the lithosphere. If the lithospheric plate is more or less uniform in the direction of compression, then its failure and the formation of a subduction zone occur through the localization of deformation in a narrow area. That process develops when the compressive stress in the plate reaches the maximum, the lithospheric yield limit σ_s. If the lithosphere contains a weak zone or inclined fault (for example, an old transform fault) oriented across the compression, then the subduction can be initiated at these inhomogeneities before compressive stresses reach the yield limit. Below we consider variants.

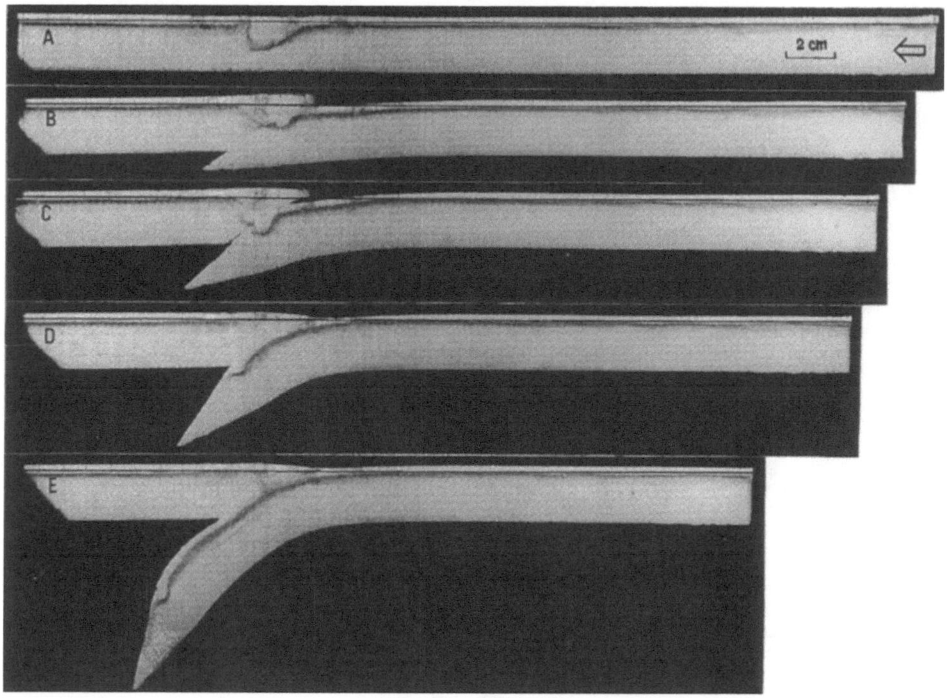

Figure 3.9. Experiment 3.8 (τ_s^m is the same as in experiments 3.1 to 3.5: τ_s^m = 15 Pa).

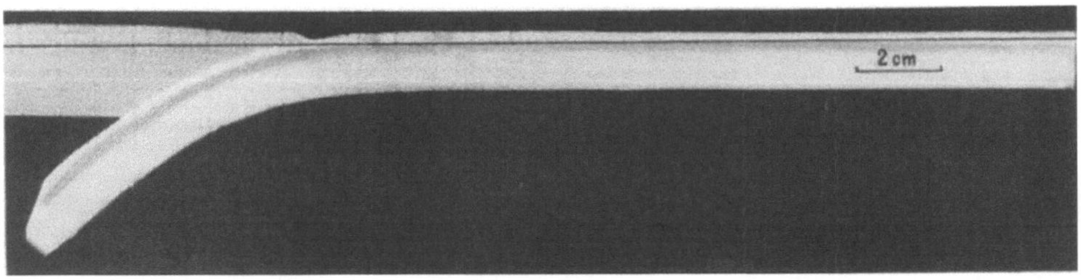

Figure 3.10. Experiment 3.9 (the subducting plate thickness H is 1.5 times smaller then in the previous experiments).

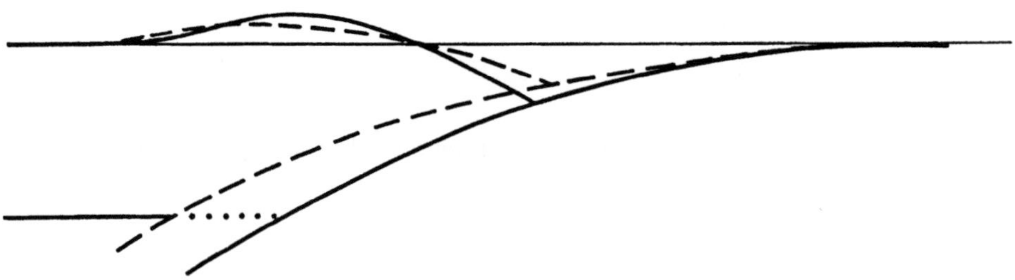

Figure 3.11. Influence of the dip angle of the interplate surface on the nonisostatic relief in a subduction zone. The thin horizontal line indicates the level of the isostatic equilibrium of the lithosphere; solid and dashed lines correspond to steeper and flatter dip angles, respectively.

3.4. Formation of a Subduction Zone in a Homogeneous Oceanic Lithosphere

In this scenario increasing horizontal compression of the lithosphere first produces thrusting in its upper layers with faults oriented perpendicular to the compression. The plate then undergoes flexural buckling and assumes a wavelike shape with a wavelength λ that depends on the lithospheric parameters H, σ_S, and E. λ value decreases with further compression. Prior to failure of the plate (localization of deformation), λ reaches average value of about $3H$, i.e., about 200 km in nature for H = 60-70 km (this minimal λ value is weakly dependent on the other parameters of the lithosphere). Afterward, the deformation is localized in one of the flexural sags (the localization may be initiated in several sags, but continues to develop in one only). The lithosphere thickens in the zone of deformation localization, with a ridge growing on the surface (Figures 3.3*f* and 3.4*b*). Two major intersecting thrust faults form during this process (Figure 3.4*b*). Complete failure of the lithosphere along one of them results in the initiation of subduction. This mechanism for the development of subduction produces a distinct step like structure of the inner trench slope (Figure 3.4*c*). The lower, steeper part of the slope comprises a strongly dislocated block of upper lithospheric rocks including the sediments, splinters of the oceanic crust, and mantle rocks. The block forms the slope break dividing the inner slope into two different parts. In nature the slope break can produce a damming effect for the sediments shedding off the arc resulting in formation of a sedimentary terrace as shown in Figure 3.4*c*. Such features, the deformed lower part of the slope, the slope break and the deep sea terrace do exist in actual subduction zones [*Karig and Sharman*, 1975] (see Figure 3.12).

It is interesting to note that in earlier works the step-like structure of inner trench slopes was attributed to accretion of the sediments scraped from the subducting plate. It was thought that the area of the slope from the trench to the slope break completely formed by stripped and accreted sediments [*Karig and Sharman*, 1975; *Dickinson and Seely*, 1979]. However, drilling data [*von Huene and Uyeda*, 1981], evidence from dredging [*Fisher and Engel*, 1969], as well as seismic profiling [*Nishenko and McCann*, 1979] have shown that accretionary prisms are often smaller or entirely absent. The frontal part of the overriding plate has almost no pelagic sediments in many subduction zones and is composed of deformed volcanoclastic rocks and blocks of oceanic crust [*Grow*, 1973; *Moore and Karig*, 1976; *Aubouin et al.*, 1982]. Such a composition is consistent with the

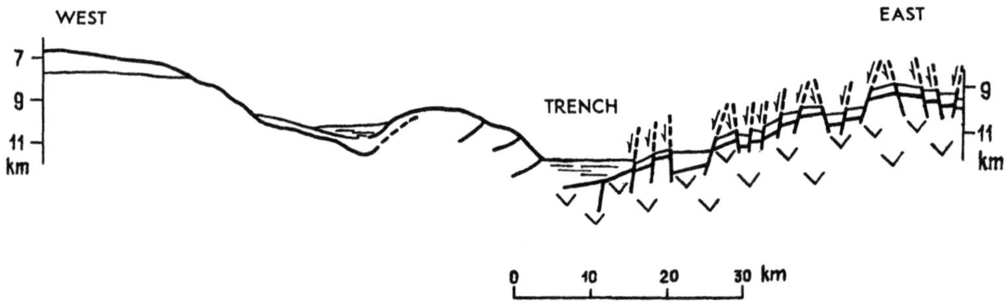

Figure 3.12. Interpreted seismic profiles across the Izu-Bonin trench [after *Ogawa et al.*, 1989].

model in Figure 3.4*c*.

The area of inner trench slope undergoes alternating vertical movements during the formation of a subduction zone. For example the slope break at first subsides during the stage of flexural buckling (Figure 3.4*a*). The subsidence is followed by uplift due to deformation localization (Figure 3.4*b*), and then by subsidence again at the stage of formation of the trench and initiation of subduction (Figure 3.4*c*). The amplitude of these oscillations in nature must amount to a few kilometers.

3.4.1. Intraplate Deformation in the Northeastern Indian Ocean

It follows from the experiments that localization of deformation in the lithosphere and formation of a subduction zone is preceded by a flexural buckling of the plate. Such buckling probably takes place in the northeastern Indian Ocean within a well known seismically active area of intraplate deformation (see Figure 3.13). The seismicity is associated with extensive young NE thrusting (Figure 3.13) [*Levchenko*, 1986]. Fault density varies over the seafloor, but is particularly great in the northern Central Basin where thrusts occur every 5-20 km [*Levchenko*, 1986]. Faulting is seen there superimposed upon the background of large-scale basement undulations (folds) with a wavelength λ of 100 to 300 km and an amplitude of 1 to 1.5 km [*Weissel et al.*, 1980], see

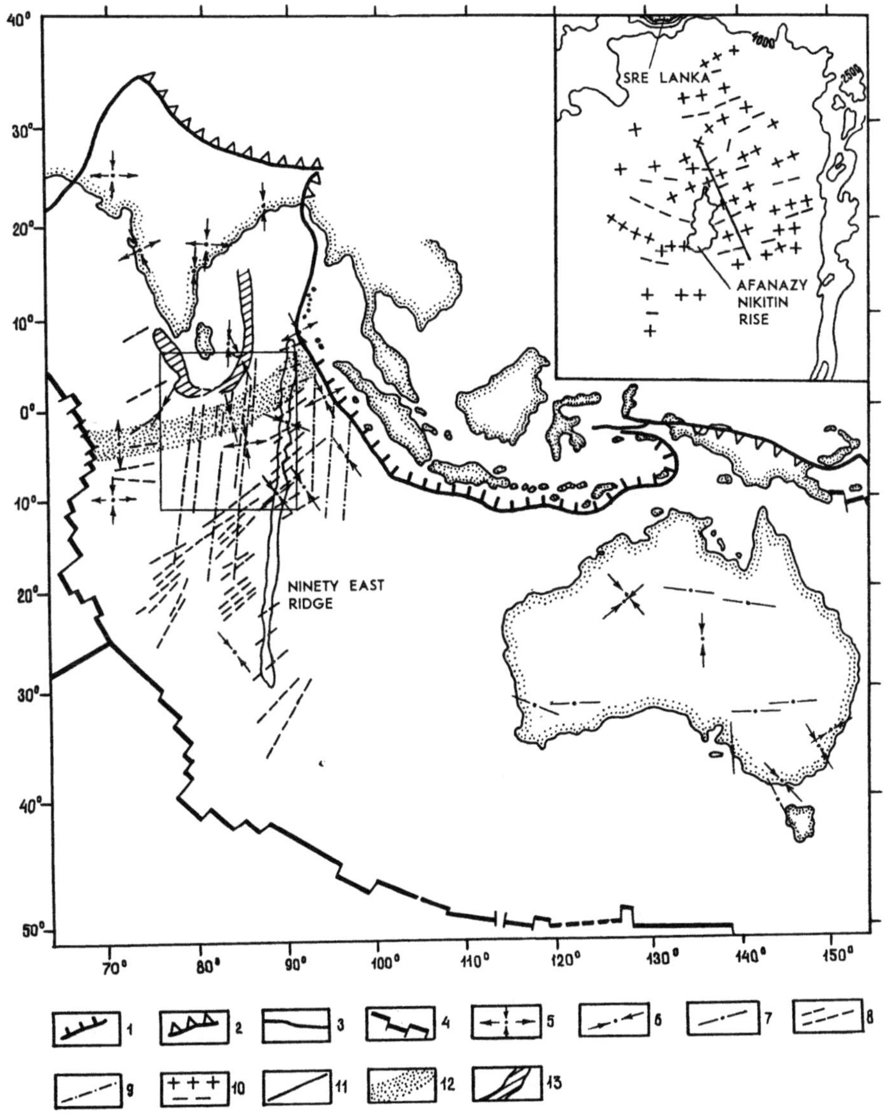

Figure 3.13. Indo-Australian plate
1, subduction zone; 2, collision zones; 3, and 4, plate boundaries (3, transform, 4, divergent) [*Weissel et al.*, 1980]; 5, epicenters of earthquakes with strike-slip focal mechanism; 6, thrust fault mechanism; 7, directions of the maximum compressional axis (from in situ measurements) [*Sykes*, 1970; *Stein and Okal*, 1978]; 8, young (or rejuvenated) faults; 9, old transform faults [*Levchenko*, 1986]; 10, basement highs (pluses) and lows (minuses) [*Weissel et al.*, 1980] (see inset); 11, location of the profile shown in Figure 3.14; 12, position of the hypothetical future subduction zone; 13, Komorine and 85 E ridges [*Levchenko*, 1986].

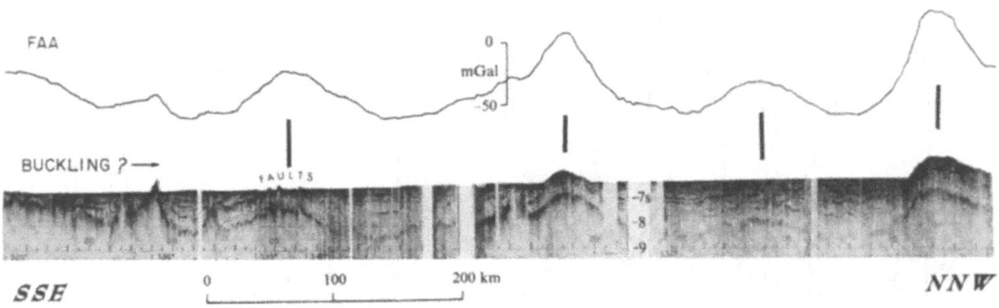

Figure 3.14. Seismic and free air gravity anomaly profiles across the intraplate deformation zone in the Central Indian Basin (the profile location is shown in the inset of Figure 3.13) [*Weissel et al.,* 1980].

Figure 3.14. Both the flexure of the basement and the faults which cut it are largely flattened by the sedimentary cover. At the same time, the long-wavelength basement topography is clearly reflected in differential free air gravity anomalies Δg which reach values of 80 mGal (Figure 3.14) [*Weissel et al.*, 1980]. Such high Δg values are difficult to explain by isostatic mechanisms such as variation of the crustal thickness. The basement topography is isostatically uncompensated, which is likely due to the flexure involving the whole lithospheric thickness.

The cause for the intraplate deformation seems to be subhorizontal northwest-southeast compression of the lithosphere inferred from seismological [*Sykes*, 1970; *Stein and Okal*, 1978; *Wiens et al.*, 1986] and geomorphic [*Weissel et al.*, 1980; *Levchenko*, 1986] observations (see Figure 3.13). This compression is believed to have been originated as a result of the India-Eurasia collision [*Weissel et al.*, 1980]. The deformation itself is considered by many authors to be a result of flexural buckling of the Indo-Australian Plate due to the compression [*Weissel et al.*, 1980; *Cloetingh and Wortel*, 1985; *Shemenda*, 1989*b*]. The conditions for such a buckling to occur have been studied by mathematical modeling [*McAdoo and Sandwell*, 1985; *Zuber*, 1987]. The experiments presented above (experiments 3.4 and 3.5, Figures 3.5 and 3.6) also corroborate the idea of flexural buckling of the Indo-Australian Plate. The average flexure wavelength λ in the model and in nature (with the scale factor taken into account) are nearly the same, about 200 km on average. (In fact, λ is not constant along the oceanic plate, neither in nature (Figure 3.14) nor in the model (Figures 3.5 and 3.6) contrary to what one would expect basing on the solution for an elastic lithosphere [*Turcotte et al.*, 1982]). However, a more careful analysis of the intraplate deformation and thrusting in the Indian Ocean raises a number of questions. In particular, the deformation seems to concentrate in the region adjoining the southern edge of Hindustan and Sri Lanka (Figure 3.13). Such a feature is not predicted by the model. In addition, it appears that the zone of intraplate deformation is far from having a simple structure of linear depressions and rises (as in experiments), but rather is spotty [*Levchenko et al.*, 1990]. These features may in part be related to the fact that simple 2-D modeling is too rough an approach to an essentially 3-D situation in the zone of intraplate deformation due to a complex geometry of the continental margin in that area. Below we present the result of 3-D experiment which takes into account to a first approximation the configuration of the Indian continental margin (i.e., the boundary between the thick and thin lithosphere) in the northern Indian Ocean and the adjacent boundaries of the Indo-Australian Plate.

Experiment 3.10. The northwestern part of the Indo-Australian Plate is a continuous layer in the model with a greater thickness under Hindustan (Figure 3.15). The plate abuts a rigid stop (bath wall) in the northwest, while the eastern and western boundaries of the plate, corresponding to the Sunda trench, the Carlsberg and Central Indian Ridges are free (the layer thickness along them is zero, the liquid "mantle" is at the surface). Compression is produced by a piston moving from the southeast at a constant rate (Figure 3.15*a*).

Similar to the previous experiments, the initial stage of deformation involves a plastic compression of the oceanic lithosphere which does not, however, develop uniformly throughout the whole plate, but with greater intensity in the region (approximately outlined by dots in Figure 3.15*a*) that abuts the Indian block from the south. The plate then buckles in this region, with two frontal folds (which have a slightly curved shape around the southern tip of the Hindustan) being most pronounced. As in experiments 3.3, 3.4, and 3.5 (Figures 3.3, 3.5, and 3.6) small reverse faults appear at the plate surface, mostly in depressions (they cannot be seen in the photographs). The deformation then is localized in some of the sags causing the formation of a set of secondary fractured rises (ridges) which complicate the pattern of initial "flexural" folding. Later, inclined major faults crossing the

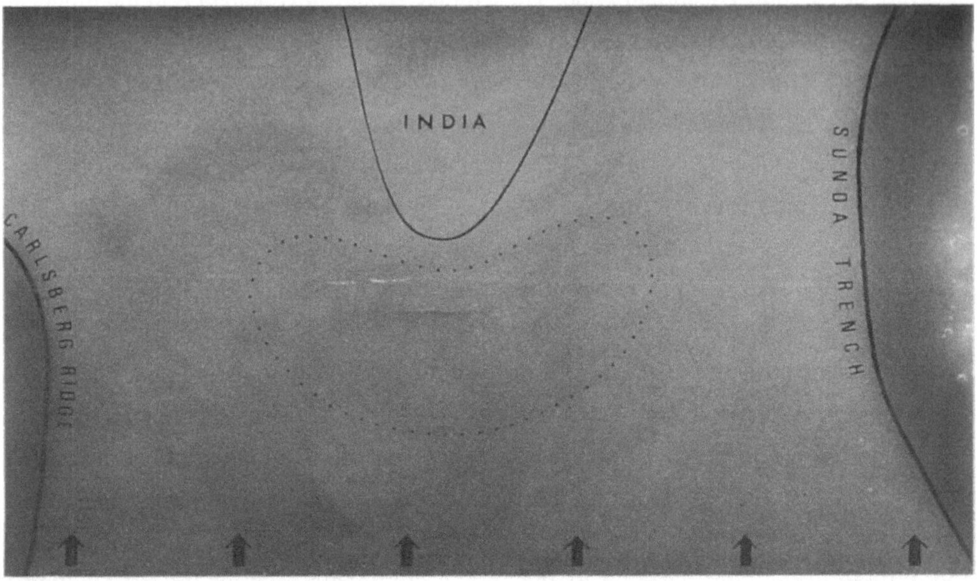

(A)

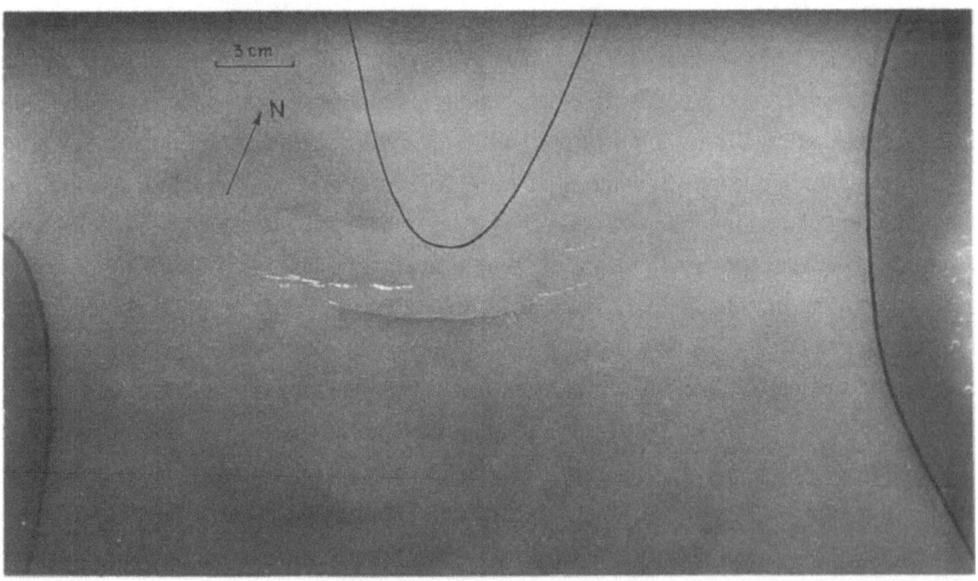

(B)

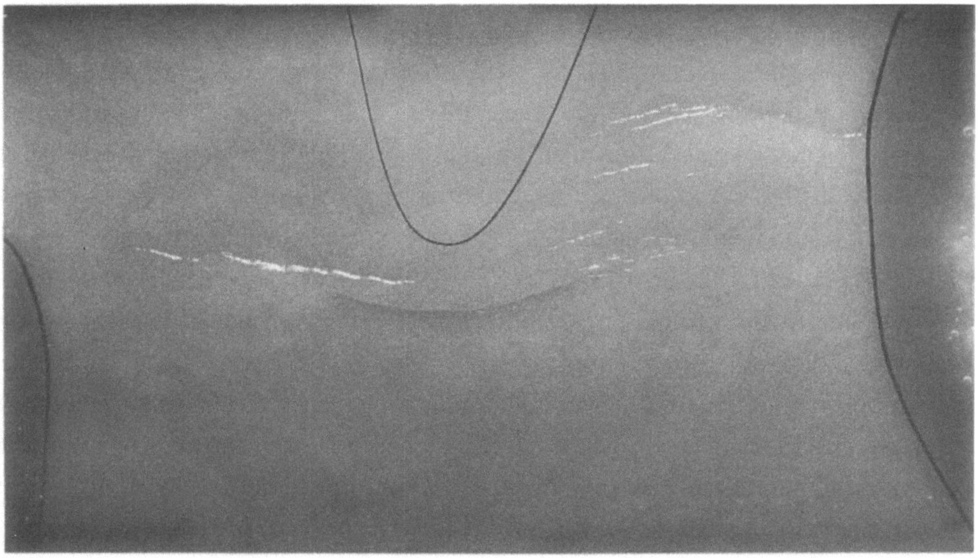

(C)

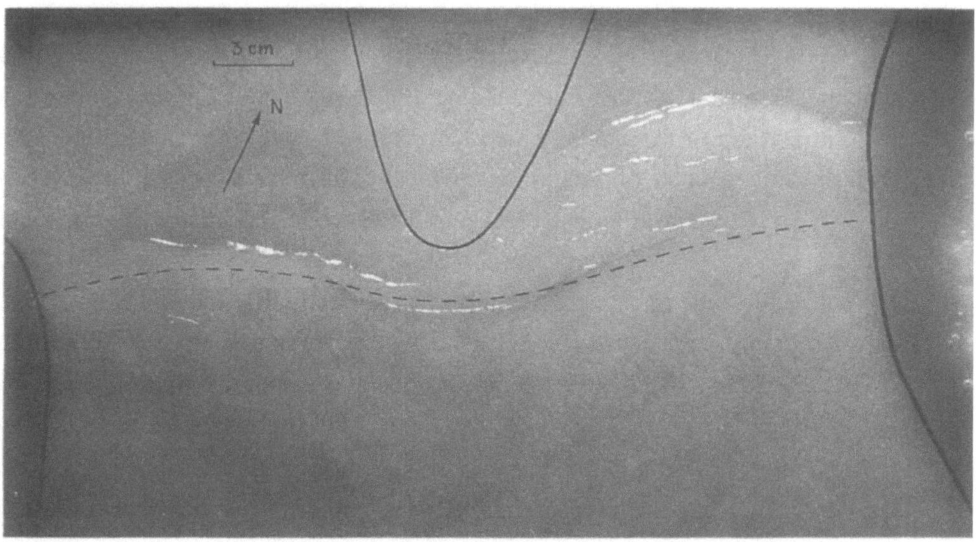

(D)

Figure 3.15. Result of experiment 3.10 (viewed from above)
(*a*) to (*d*) Stages of the model process; dark and light narrow lines on the photographs represent reverse faults crossing the entire lithosphere thickness and dipping northward and southward, respectively; the dotted line in Figure 3.15*a* approximately outlines the area of most intensive plastic deformation at the initial stage of the experiment; dashed line in Figure 3.15*d* shows the ultimate position of the forming subduction zone.

whole lithospheric thickness form in the zones of deformation localization. In contrast to the preceding experiments, these faults occur at different places (different downwarps) and can dip in opposite directions. Figures 3.15*a* and 3.15*b* show two such faults which are located on the opposite sides of the "flexural" rise and dip in opposite directions (see caption to Figure 3.15). Such "irregular" location of zones of deformation localization and major faults in the lithosphere is typical of three-dimensional models. Further compression produces thrust (underthrust) displacements along the faults, with the faults themselves propagating laterally (Figure 3.15*b*). The slab flexure occurs on both sides of the initial zone of deformation concentration (between this zone and the lateral plate boundaries). The folds which develop there are perpendicular to the axis of compression and are not conformable with the trend of "older" folds in the central zone. The buckling and thrust-strike slip faults occur then off the eastern side of the Hindustan (Figure 3.15*c*) as well as on the sides of the central zone, conforming to the trend of the folds already formed there. These faults interact with those propagating out of the central zone and produce the ultimate shape of a new convergent plate boundary with a configuration outlined approximately by a dashed line in Figure 3.15*d*. This boundary did not form in one event as a single unit zone of one-directional subduction but resulted from a sequence of restructurings of local subduction zones that formed at individual faults and had different polarities.

It should be noted that the main sequence of events described above was repeated from experiment to experiment. At the same time, the details of the model process such as the exact position and dip of the faults, the sequence of restructuring of local subduction zones, as well as the polarity of the ultimate subduction may have been different. These details are very sensitive to various heterogeneities in the model, either random or made on purpose.

The presented 3-D experiments provide further confirmation for the model of flexural buckling of the Indo-Australian Plate and clarify some details in the structure of the deformation zone. In particular, the modeling corroborates the intuitively understandable cause for deformation concentration in the northern Central Basin as being due to the southern promontory of India jutting out into the thinner oceanic lithosphere. Also, the model provides insight into the possible causes for the complex irregular (spotty) structure of the deformation zone. It may be caused by the formation of two types of rises, those due to the lithospheric flexure (flexural rises) and those due to the localization of deformation and thrusting. Certain strongly dislocated basement highs seen on seismic profiles through the zone of deformation can be formed due to the deformation localization.

Such highs develop in "flexural" depressions (see Figure 3.4*b*). Their position is determined by different lithospheric heterogeneities. For instance, the old north-south transform faults shown in Figure 3.13 which "cut" the lithosphere into autonomous bands must play an important role. The evolution of deformation within different bands may develop independently, resulting in offset of topographic features that form due to deformation.

3.4.2. Polarity of Subduction Zones, Trapped Back Arc Basins

Experiments 3.3, 3.4 and 3.5 (Figures 3.3, 3.5, and 3.6) show that failure in the lithosphere under horizontal compression does not occur directly in the continental margin (even though there are subvertical faults through the entire plate), but at some distance toward the ocean. This rather surprising result is due to the buckling of the oceanic plate prior to failure. The wavelength of the buckling λ controls the distance from the continental margin to the location of lithosphere failure. We conducted experiments in which bending of the oceanic slab was rendered impossible due to its small length (the length was made less than $\lambda/2$). Failure in the oceanic lithosphere occurred directly at the boundary between the thick and thin plates (or near the piston) through the formation of the major thrust fault dipping either under the continent or under the piston. Accordingly, subduction also started beneath the continent (piston).

In experiments 3.3, 3.4 and 3.5 the lithosphere fails along one of two intersecting faults dipping in opposite directions. In experiment 3.3 (Figure 3.3) the final fault dips under the continent, while in experiments 3.4 and 3.5 (Figures 3.5 and 3.6) it dips seaward. It has not been reliably established which direction is the preferred one. The next experiment demonstrates that probably both possibilities are equally likely.

Experiment 3.11. The conditions of this experiment are similar to experiment 3.3 (Figure 3.3), except that the plate has first been separated into two halves by a longitudinal vertical cut (see Figure 3.16). The initial stages of experiment 3.11 are also similar to experiment 3.3. The oceanic lithosphere bends, with thrusting developing on the surface. Then a zone of deformation localization common to both halves with two major thrust faults starts to develop as usual. However, complete failure of the lithosphere in the two halves occurs along different faults dipping in opposite directions. Subduction develops in opposite directions too, resulting in gradual offset of subduction zones on both sides of the initial cut.

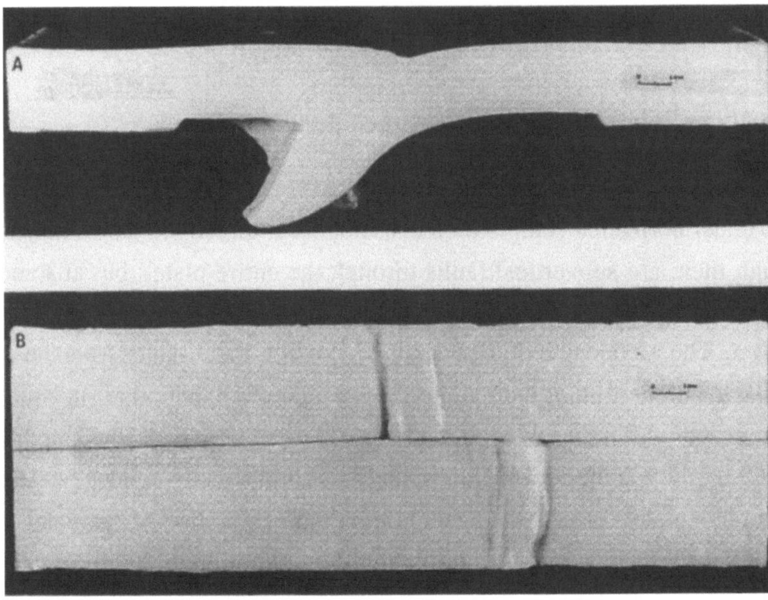

Figure 3.16. Experiment 3.11; the conditions are the same as in experiment 3.3 (Figure 3.3), except that the model lithosphere had first been separated with a longitudinal vertical cut into two halves
(*a*) Section of the frozen deformed model; (*b*) Deformed model as viewed from above.

Failure of the lithosphere along one or the other directions has important geotectonic consequences. For example, if the surface of the plate fracture dips under the ocean, as in experiments 3.4 and 3.5 (Figures 3.5 and 3.6), then subduction of the oceanic plate segment located between the subduction zone and the continental margin inevitably would result in their collision (in the obduction of the oceanic plate on the continental lithosphere). Further compression may produce failure of the oceanic lithosphere in a new place followed by the jump of subduction zone (see Chapter 5). One of the possibilities is failure of the oceanic lithosphere directly at the continental margin, with subduction being initiated under the continent. Such an episode may have occurred during the evolution of the

Andean margin of South America where subduction currently occurs directly beneath the continent.

If, on the other hand, plate failure in the zone of deformation localization occurs along the major fault dipping under the continent from the very beginning (as in experiment 3.3, Figure 3.3), then the subduction is stable. The marginal basin trapped between the continent and the subduction zone can persist for a long time. The dimensions of the basin are controlled by several factors. In the simplest 2-D situation without sedimentary load at the continental margin (experiments 3.3 and 3.4, Figures 3.3 and 3.5), the basin width d is half the flexural wavelength λ, i.e. is of the order of 100 km. The presence of a sedimentary wedge in the continental margin may significantly increase d (experiment 3.5, Figure 3.6). The width of the marginal sea is largely controlled by the wedge parameters. It was very difficult to determine quantitative relationship between d and the wedge parameters in the experiments (this is a problem for the mathematical modeling). Only some qualitative patterns have been catched more or less definitely. For example, if the height of the wedge base h_s is smaller than some value h_0, then (as was mentioned above) such a wedge practically does not affect the position of the zone of deformation localization and failure in the lithosphere. If, on the contrary, $h_s > h_0$, and the wedge width d_s does not exceed some value d_0, the localization of deformation occurs at a distance of about $\lambda/2$ or somewhat farther from the wedge apex (as in experiment 3.5, Figure 3.6). The plate is convex everywhere beneath the wedge, and hence has the same sign of curvature. When $h_s > h_0$ and $d_s > d_0$, the slab may assume a wavy shape directly under the wedge. Failure in the lithosphere under these conditions occurs closer to the continent than for a smaller d_s. These patterns could be probably studied in the zone of intraplate deformation in the Indian Ocean. This zone at the area where the lithosphere is cut by old north-south faults into bands (Figure 3.13) and is under the load of a thick sedimentary wedge (the parameters of the wedge vary from band to band) may be a convenient natural test site.

One more factor determining the dimensions and shape of the trapped basin is the configuration of the continental margin. Experiment 3.10 (Figure 3.15) demonstrated that subduction fist initiates near the protrusion of the continental lithosphere into the ocean. Local subduction zones then propagate from the protrusions along the shortest route to meet each other. The resulting subduction zone does not mimic the configuration of the continental margin and separate oceanic basins between the continental protrusions.

Natural examples of trapped basins are the Aleutian Basin in the Bering Sea, the Sea of Okhotsk, excluding the Kuril Trough and some others.

3.4.3. Application of the Model to Oceanic Transform Faults

A small convergent or divergent component of motion is always present in oceanic transform faults along with the main strike slip component. The sign and magnitude of that small component largely determine the relief and deep structure of the transform fault [*Ushakov et al.*, 1979; *Dubinin*, 1987]. The model in Figures 3.4*a* and 3.4*b* can formally be applied to the faults with transverse compression. Indeed, lithospheric blocks of different ages and hence of different thicknesses contact along a transform fault which is supposed to be nearly vertical. If the fault is regarded as a two-dimensional feature, then strike-slip displacement along it does not produce lithospheric deformation. Deformation occurs only due to the transversal compression according to the model in Figures 3.4. The model predicts a formation of the ridge due to the localization of lithospheric deformation and thrusting in the crust at a distance of about 100 km ($\lambda/2$) from the fault. The ridge can be location of the future subduction zone. Although it would be interesting to verify this prediction thoroughly, it is however, beyond the scope of this work. We only note that some transform faults characterized by transverse compression are indeed associated with ridges which can be regarded to have been formed due to the localization of a lithospheric (crustal) deformation. One possible example is the Gorrindge Bank associated with the eastern Azores-Gibraltar Fault [*Verzhbitsky et al.*, 1989] and the seamounts of the Hosshu system located about 100 km to the south. Some authors propose that northward subduction is already developing under the Gorrindge Bank [*Le Pichon et al.*, 1970]. Another example is the northeastern segment of the Owen transform fault in the Indian Ocean. There is also a rise associated with the crustal thickening to the northwest of the fault at a distance of about 100 km from the fault [*Whitmarsh*, 1979].

To summarize, it seems that analogues for all stages of subduction zone formation considered in this section can be found in nature. The initial stage, the flexural buckling of the oceanic lithosphere (Figure 3.4*a*), appears to take place in the northeastern Indian Ocean, the intermediate stage, the localization of lithospheric deformation (Figure 3.4*b*), applies to the Gorrindge Bank, and the final stage, the subduction is represented by subduction zones. Similar processes can develop also within the continental lithosphere. Localization of deformation under strong compression can occur in the depressions caused either by flexural buckling of the lithosphere with inhomogeneous thickness or by the load of sediments within the sedimentary basins. Crustal thrusting in the zone of deformation localization will produce ridge. If compression is large enough this process will be followed by continental subduction which is modeled in Chapter 6.

3.5. Initiation of a Subduction Zone on an Old Inclined Fault

There is another "easier" way for subduction zone to form when the lithosphere contains an inclined old fault zone oriented transverse to the compression axis. Subduction initiates at the fault sooner than the compressive stress reaches the value sufficient to produce flexural buckling of the lithosphere. The inner slope of the trench forming during initial stages of subduction has a structure that is much different from the preceding model in Figure 3.4*c*. The structure is largely controlled by a deformation of the frontal edge of the overriding slab when the edge descends onto the surface of the subducting lithosphere. This effect can be seen in experiments 3.6, 3.7 and 3.8 (Figures 3.7, 3.8 and 3.9). The frontal parts of the originally horizontal overriding slab subside at the beginning of these experiments forming the inner trench slope. The subsidence develops under the weight of the frontal parts of the overriding plate when the underlying lithosphere bends down. If the plate is too strong (as in experiment 3.6, Figure 3.7) then the weight is not sufficient to overcome the lithospheric yield limit τ_s and to produce the subsidence. In experiments 3.7 and 3.8 (Figures 3.8 and 3.9) the lithosphere is weaker and the frontal wedge of the overriding plate experiences plastic deformation subsiding under its own weight. The subsidence is accommodated in both normal faulting in the overriding wedge (shown in Figure 3.17) and its bending. The details of these deformations are not seen in the above experiments due to the small size of the models. Therefore we present below another simpler experiment which illustrates the effect in question more clearly.

Experiment 3.12. A horizontal layer that possesses dilatant-plastic properties (clay) rests on a rigid base consisting of two parts separated by an inclined surface (Figure 3.18). Subsidence of one of the parts (which imitate the subsidence of the subducting plate) produces bending of the layer (with tension in the upper horizons) and normal fault dipping under the "trench".

The formation of the subduction zone at the preexisting inclined fault thus results in extensional tectonics of the inner trench slope contrary to the mechanism described in the Section 3.4 which produces compressional tectonics. Both types of inner trench slope tectonics can be found in nature. The Mariana [*Hussong and Uyeda*, 1982], Tonga-Ker-

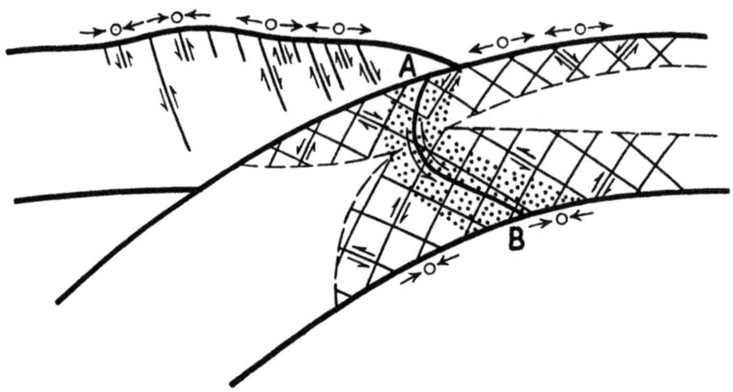

Figure 3.17. Schemes for the overriding wedge (plate) deformation (for subduction initiating at preexisting inclined fault) and for the quasi-steady deformation of the subducting plate (see explanation in the text).

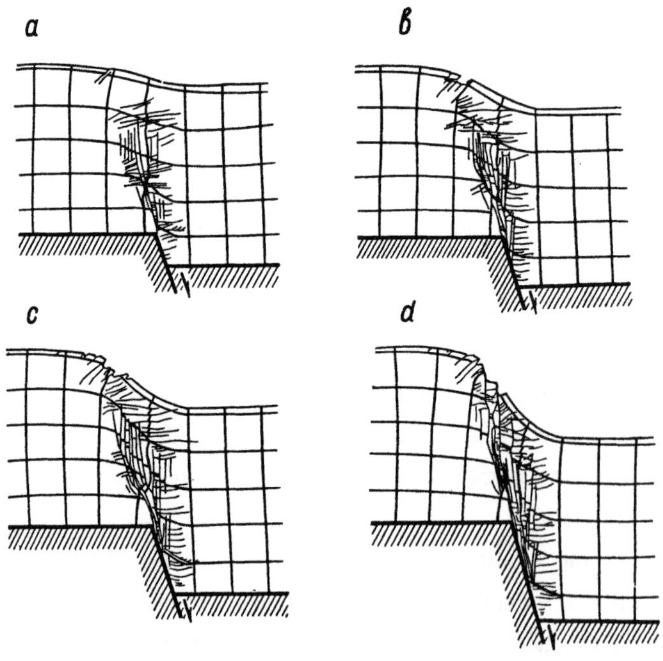

Figure 3.18. Experiment 3.12; deformation of the clay layer due to subsidence of one part of the rigid base relative to another (performed by A. Bokun).

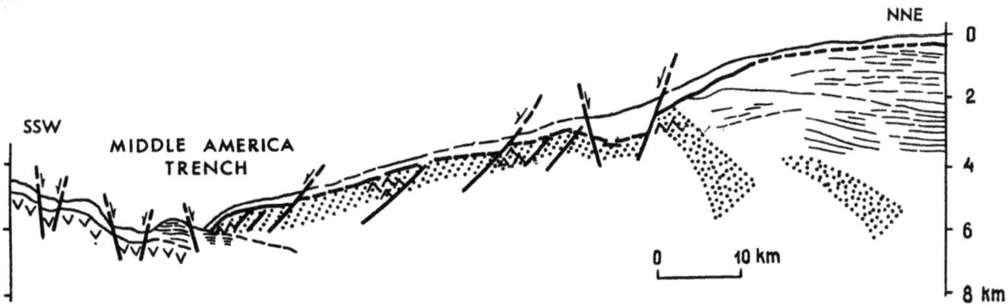

Figure 3.19. Schematic section through the Guatemala trench [after *Aubouin et al.*, 1982].

madec [*Gnibidenko et al.*, 1985], and Central American (Figure 3.19) trenches are examples with tension within the inner trench slope.

Tension develops only in the most frontal parts of the overriding wedge undergoing subsidence during initial subduction. Farther up the inner slope, the wedge is uplifted forming an isostatically uncompensated frontal arc. The uplift also causes plastic deformation in the wedge resulting in formation of reverse faults as shown in Figure 3.17. The structure of the overriding wedge in this figure forms in the initial stage of subduction. It can be considerably changed during further evolution of subduction due to various factors. For example, friction between the plates can produce reverse motion along the normal faults [*Lobkovsky et al.*, 1980]. Still greater changes can be associated with a change in the interplate dip angle ψ, due for example, to a nonuniform erosion of material at the base of the overriding wedge. If the angle decreases then according to Figure 3.11, the fore arc region subsides and the most frontal portion of the overriding wedge is uplifted. The direction of movement along the faults in the overriding slab shown in Figure 3.17 will be reversed during this process. Similar effects can also occur due to the change in nonhydrostatic pressure σ_n between the plates when subduction mode changes from Chilean- to Marianas-type and vice versa (see Section 4.12).

3.6. Strain and Stress in the Subducting Plate

While the greatest deformation of the overriding wedge develops during initiation of subduction, the subducting plate undergoes large quasi-stationary deformation all the time. This deformation was studied using the technique for visualization of deformation described in Section 2.4. The results have shown that subducting plate experiences flexure in front of the subduction zone, leading to compression of the material at the base of the plate and extension at the top (Figure 3.17). Normal nonhydrostatic horizontal stresses σ_h in the upper and lower layers of the plate exceed the lithospheric yield limit resulting in plastic deformation. The area of plastic deformation includes the outer (oceanic) trench slope and part of the outer rise. As the average stress σ_h through the plate thickness is compressive (is of about $0.3\sigma_s$, see Section 3.2), the lower plastic zone is thicker than the upper one (Figure 3.17). The curvature of the subducting plate increases toward the trench (see Section 3.8 and Figure 3.7) which causes an increase in deformation and widening of both lower and upper plastic zones (Figure 3.17). On the model surface the deformation is manifested in small linear normal faults forming within the outer trench slope parallel to the trench. Closer to the area of maximum curvature of the plate which during mature subduction is somewhere under the frontal segments of the overriding plate (a stippled area in Figure 3.17) the deformation of the subducting plate becomes more complex. Thickening both upper and lower plastic zones are merged and connected by slip lines which accommodate shear (thrust) movements of the material. When this area has been passed, the thickness of the plastic zones and the deformation itself decrease sharply. The lithosphere is sinking further into the mantle without deformation, since the plate is in hydrostatic equilibrium with the surrounding "mantle" (their densities are equal) and does not undergo any dynamic influence from it. If there is no equilibrium, the deformation pattern would apparently change.

The comparatively sharp change in the direction of the subducting plate motion in the subduction zone occurs thus due to the superposition of the simple bending of the plate and shear (thrust) deformation along the zone dipping under the trench. (Existence of such a zone has been supposed by *Lobkovsky and Sorokhtin* [1976]). The value of the shear and its contribution to the total deformation of the lithosphere depend on the dip angle, ψ, of the interplate surface. This deformation decreases with decreasing ψ and the shear zone becomes less distinct. Finally, the zone disappears completely. Increasing ψ as well as decreasing τ_s makes the shear deformation more intensive and the shear zone narrower and well pronounced. Below we present the result of the experiment with large ψ value. The subducting plate experiences large shear deformation, with the "bending component" being very small.

Experiment 3.13. The overriding plate is replaced by a rigid stop dipping at an angle of 45°, the subducting plate is the same as in the previous experiments. This plate undergoes a sharp flexure in front of the stop during subduction. Its deformation produces a significant increase in plate thickness (Figure 3.20) (lesser thickening was also observed in the preceding experiments). The original angle of 90° between the transverse lines drawn

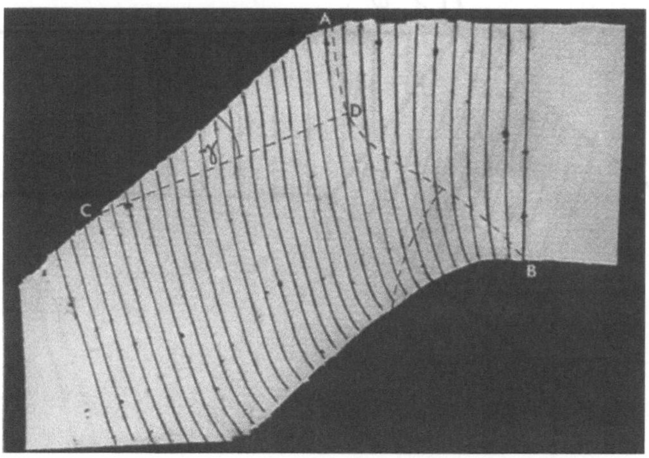

Figure 3.20. Experiment 3.13; subduction under a rigid stop dipping at an angle of 45°.
The photograph shows the part of the deformed (during mature subduction) and frozen plate model in the subduction zone; the transverse marker lines in the plate had originally been vertical and perpendicular to the surface of the undeformed specimen; dashed lines show the slip lines (see the text for more explanations).

on the lithospheric model and its surface decreases when the slab passes the inflexion zone. The character of curvature of the lines indicates that the dominant type of the deformation is shear along the line (zone) AB (Figure 3.20). To see this one can compare the experimental picture (Figure 3.20) with the ideal scheme in Figure 3.21 derived by simple geometrical constructions. Although the patterns are similar to a first approximation, the experimental picture is more complicated. The shear zone in Figure 3.20 is not straight. Yet, practically the entire flexure area involves comparatively small deformations with more intensive deformation in the lower part of the plate and under the stop (Figure 3.20). If there is no deformation, then the lines drawn on the slab would have been parallel. In fact, there is some lag of the upper parts of the lines under the stop relative to the lower parts. The lag decreases when approaching the end of the stop (point C in Figure 3.20) and the lines become nearly parallel to each other further down along the plate. The lag zone can be fit approximately by the triangle ACD with a side abutting the

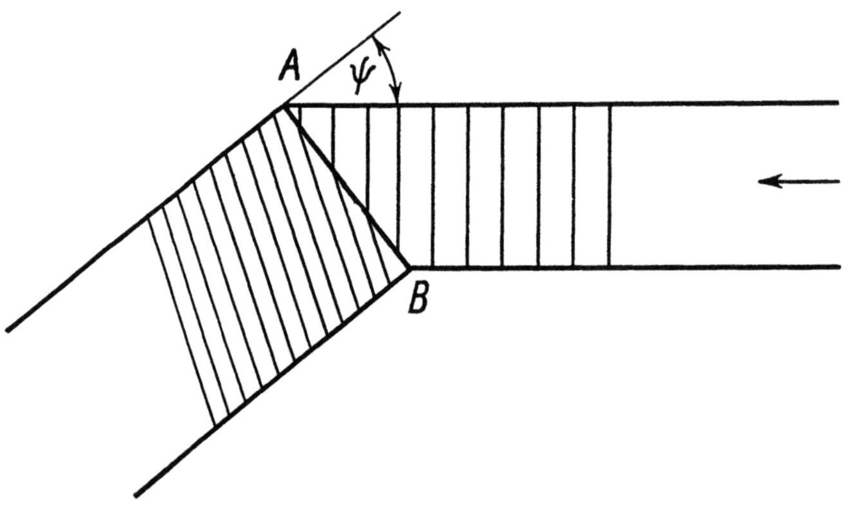

Figure 3.21. Geometry of the subducting plate and bending of the vertical markers for the ideal case of pure shear along the straight line AB.

stop surface. The formation of this zone is caused by the pressure from the stop (the overriding plate). The upper layers of the subducting plate behave as if pressed into the body of the rest of the specimen. Also, the friction exerted by the stop acts as a brake on the slab top producing further lag of the material beneath the stop. The friction value is related to the angle γ shown in Figure 3.20 (see next section). The position of the zone ACD and the character of curvature of the marker lines suggest that the zone is probably under uniform stress (i.e., the slip lines are straight there). Another mentioned zone of deformation in the lower part of the inflexion area is also triangular in shape and is caused by the flexure and compression dominating the slab base.

When subduction develops at an even steeper angle than in experiment 3.13, the shear zone within the subducting plate does indeed degenerate into a straight line as in the ideal scheme in Figure 3.21. This case is illustrated by experiment 3.1 in Figure 3.1. According to the experimental results, it seems that at realistic ψ values the shear (thrust) deformation in question is not large. It can be supposed however that other factors that have not been tested in the modeling can lead to the formation of distinct shear zone even at small ψ. For example tentative experiments have shown that the shear deformation increases and shear zone narrows with increase in brittleness of the subducting plate and/or decrease in τ_s at ψ = constant.

3.7. Estimation of Pressure between the Plates in the Subduction Zone

In this section we use the strain pattern in the subducting plate derived from experiment 3.13 (Figure 3.20) to estimate the nonisostatic pressure σ_n which the subducting plate exerts on the overriding plate [*Shemenda*, 1979; *Lobkovsky and Shemenda*, 1981]. This pressure controls the departure of the lithosphere from isostatic equilibrium and hence the anomalous gravity field in the subduction zone. Unfortunately, the pressure cannot be directly measured instrumentally, since it is very small in the model.

Taking into account the conditions of the experiments and the properties of the model lithosphere, we assume that the picture in Figure 3.20 corresponds to steady state plain strain, while the stresses in the inflexion region (plastic deformation zone) satisfy Mises-Hencky yield condition

$$\bar{\tau} = 1/2[(\sigma_{xx}-\sigma_{zz})^2 + 4\tau_{xz}^2]^{1/2} = \tau_s$$

where $\bar{\tau}$ is the intensity of deviatoric stresses in the plate; *x, z* are Cartesian coordinates with directions as shown in Figure 3.22 (the x-axis coincides with level of isostatic equilibrium); τ_{xz} is the tangential stress; σ_{xx}, σ_{zz} are normal stresses along the appropriate axes; τ_s is the shear yield limit. In addition, the stresses should satisfy the ordinary equations of equilibrium

$$\partial\sigma_{xx}/\partial x + \partial\tau_{xz}/\partial z = 0;$$
$$\partial\sigma_{zz}/\partial z + \partial\tau_{xz}/\partial x - \rho_l g = 0. \quad (3.1)$$

The stresses σ_{xx}, σ_{zz}, τ_{xz} are related with the principal stresses σ_1 and σ_2 ($\sigma_1 > \sigma_2$) as [*Kachanov*, 1969]

$$\sigma_{xx} = (\sigma_1 + \sigma_2)/2 + [(\sigma_1 - \sigma_2)/2]\cos 2(1,x)$$
$$\sigma_{zz} = (\sigma_1 + \sigma_2)/2 - [(\sigma_1 - \sigma_2)/2]\cos 2(1,x) \quad (3.2)$$
$$\tau_{xz} = [(\sigma_1 - \sigma_2)/2]\sin 2(1,x)$$

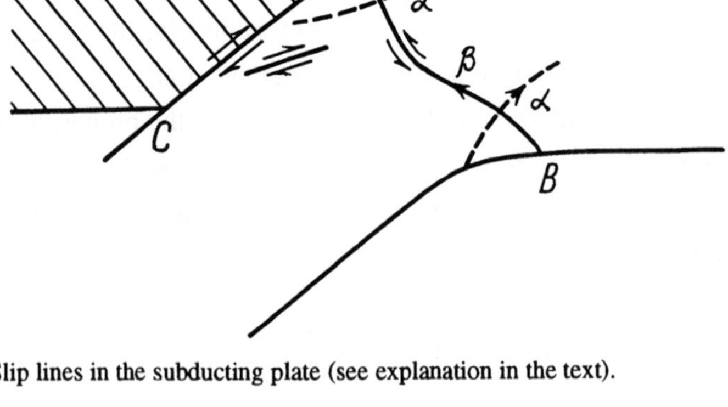

Figure 3.22. Slip lines in the subducting plate (see explanation in the text).

where $(1,x)$ is the angle between the first principal direction and the x-axis. Remembering that the relation $(\sigma_1 - \sigma_2)/2 = \tau_s$ holds in the plastic zone and defining the angle $\theta = (1,x) - \pi/4$ between the plane with the maximal tangential stress and the x-axis, one can use (3.1) and (3.2) to derive the following equations

$$\partial\sigma_0'/\partial x - 2\tau_s(\cos 2\theta\ \partial\theta/\partial x + \sin 2\theta\ \partial\theta/\partial z) = 0$$

$$\partial\sigma_0'/\partial z - 2\tau_s(\sin 2\theta\ \partial\theta/\partial x + \cos 2\theta\ \partial\theta/\partial z) = 0, \qquad (3.3)$$

where $\sigma_0' = \sigma_0 - \rho_l g z$; $\sigma_0 = (\sigma_{xx} + \sigma_{zz})/2$ is mean pressure. The relationships (3.3) are hyperbolic equations in the variables σ_0' and θ. The definition of θ implies that the orthogonal families of characteristics for equations (3.3) coincides with the slip lines (i.e., lines of maximal tangential stresses τ_s). The quantity σ_0' corresponds to the mean nonhydrostatic pressure.

The variables σ_0' and θ satisfy the Hencky relations along the characteristics of (3.3) [*Kachanov*, 1969]:

along the lines of one family of characteristics (family α)

$$dz/dx = \mathrm{tg}\ \theta;\quad \sigma_0'/2\tau_s - \theta = \xi$$

and along the lines of the other family, β, (3.4)

$$dz/dx = -\mathrm{ctg}\ \theta;\quad \sigma_0'/2\tau_s + \theta = \omega$$

where ξ and ω are parameters that characterize specific slip lines from families α and β, respectively.

To determine the pressure between the plates, it is sufficient to find the pressure at any point on the line AC (for instance, at point A) as the stress-state in the triangle ACD is uniform. It has been shown above that large shear deformation in the inflection region occurs approximately along the line AB which obviously is a slip line. For this reason it is sufficient to find the stress at any point on AB and then to determine stresses at A using equations (3.4).

Due to the very low viscosity of the asthenosphere, the tangential stresses at the base of the lithosphere are equal to zero. Hence the angle of intersection between AB and the plate base must be equal to $\pi/4$. The pressure at point B is equal to $P_B = \rho_a/gz_B$ where ρ_a is density of the asthenosphere ($\rho_a \approx \rho_l$) and z_B is the vertical coordinate of B. This pressure can be also written as [*Kachanov*, 1969]

$$-P_B = \sigma_B - \tau_s \sin 2(\theta - \chi), \tag{3.5}$$

where σ_B is the mean pressure at B; χ is the angle between the normal to the plate base and the x-axis. Passing to nonhydrostatic stresses, i.e., subtracting $\rho_l g z_B$ from both sides of (3.5), we get

$$\sigma'_B = \tau_s \sin 2(\theta - \chi) \tag{3.6}$$

where σ'_B is the mean nonhydrostatic pressure at point B. To determine the σ'_B value one should find the direction of the slip line α. This can be done from the following considerations. The distance between transverse markers in the lower part of the inflection zone in Figure 3.20 decreases during subduction due to some flexural component in the deformation in that region. Consequently, the following inequality must hold at B

$$\sigma_t^B > \sigma_n^B \tag{3.7}$$

where σ_t^B and σ_n^B are the normal stresses at B, parallel and orthogonal to the plate base. These stresses can in turn be written in the form

$$\sigma_t^B = \sigma_B + \tau_s \sin 2(\theta - \chi) \; ; \; \sigma_n^B = \sigma_B - \tau_s \sin 2(\theta - \chi) \tag{3.8}$$

Substituting (3.8) into (3.7) and using the fact that $\sigma_B < 0$, we get

$$\sin 2(\theta - \chi) < -\sin 2(\theta - \chi) \tag{3.9}$$

Remembering that $\chi = -\pi/2$ at point B, inequality (3.9) will be satisfied if the direction of the slip line α is chosen as shown in Figure 3.22. Equation (3.6) can then be used to determine the mean nonhydrostatic pressure at point B (for that point we have $\chi = -\pi/2$; θ

$= \pi/4$)

$$\sigma'_B \approx -\tau_s$$

The slip line AB belongs to the β family. Therefore, ω should keep the same value along this line (see (3.4)). At A and B, ω can be written respectively, as

$$\omega_A = \sigma'_A/2\tau_s + \theta_A\,; \quad \omega_B = \sigma'_B/2\tau_s + \theta_B\,. \tag{3.10}$$

The calculations below are based on the configuration of AB shown in Figure 3.20. The angles θ_A and θ_B at A and B have the following approximate values: $\theta_A \approx \pi/9$ and $\theta_B \approx \pi/4$. Equating ω_A and ω_B from (3.10) and substituting the values indicated above for σ'_B, as well as for θ_A, θ_B, we get: $\sigma'_A \approx -0.1\tau_s$.

The contact nonhydrostatic pressure σ_n on the line AC at point A can now be determined as

$$\sigma_n \approx \sigma'_A - \tau_s \sin 2(\theta_A - \chi_A), \tag{3.11}$$

where χ_A is the angle between the normal to AC and the *x*-axis ($\chi \approx 134^o$). Substituting the values for σ'_A, θ_A and χ_A in (3.11), we get

$$\sigma_n \approx -0.9\tau_s,$$

which can be adopted as an estimate for the contact nonhydrostatic pressure at the plate boundary. The full pressure P_n at each point of AC can be derived by adding the term $\rho_l gz$ to σ_n (z is the vertical coordinate of the point). The mean pressure $\rho_l gz$ corresponds to the hydrostatic equilibrium at depth z. The excess pressure σ_n disturbs the equilibrium, producing some nonisostatic uplift of the overlying layers (those above the interplate surface). On the other hand, the friction between the plates, τ_n, pulls the overriding plate downward. The net vertical pressure σ_v is

$$\sigma_v = \sigma_n + \tau_n \mathrm{tg}\psi.$$

The friction τ_n in experiment 3.13 can be estimated from the value of γ, the angle which characterizes the upper zone of uniform stress-state in Figure 3.20, as [*Hill*, 1950]

$$\tau_n = \tau_s \cos 2\gamma .$$

Assuming the pressure σ_v to be completely spent on producing an uplift of the lithospheric material directly above the line AC (Figures 3.20 and 3.22), one can determine the uplift amplitude Δh_n

$$\Delta h_n = -\sigma_v/\rho_l g = \tau_s(0.9 - \cos 2\gamma \operatorname{tg}\psi)/\rho_l g .$$

Substituting the values $\tau_s = 1.3 \times 10^8$ Pa, $\psi = 42^o$, $\gamma = 22^o$ (see Figure 3.20), we get $\Delta h_n \approx 1$ km. The non-isostatically uplifted layer of thickness Δh_n produces a free air gravity anomaly Δg which can be estimated using the formula for a plane parallel layer, $\Delta g = 2\pi G\rho_l \Delta h_n = 140$ mGal, where $G = 6.67 \times 10^{-7}$ m/(kg $\times$ s^2) is the gravitational constant. If the interplate friction τ_n in the subduction zone were absent ($\gamma = \pi/4$) the gravity anomaly would reach the value of $\Delta g = 470$ mGal. For the actual subduction zones typical are anomalies of 150 to 300 mGal.

3.8. Elastico-Plastic Bending of the Subducting Plate, Its Yield Limit and Modulus of Elasticity

The excess pressure between the plates leads on the one hand to a non-isostatic uplift of frontal parts of the overriding lithosphere and on the other hand to a downwarping of the subducting plate and formation of a deep sea trench. Due to the elastic properties of the lithosphere and the presence of an effectively liquid substratum beneath it, bending of the subducting plate directly in a subduction zone causes deformation of this plate hundreds of km seaward, creating an extended rise of small amplitude, the outer rise. This subduction-related phenomenon is well studied and there is no doubt that both the trench and the outer rise are caused by quasi-elastic bending of the subducting plate. *Gunn* [1947] was the first who suggested a mathematical model for this bending. Numerous later studies were largely concerned with refinements of his model in order to derive analytic or numerical solutions that better corresponded to bathymetric profiles across various subduction zones. Such solutions were used to estimate parameters of the lithosphere.

It should be noted that mathematical modeling of quasi-elastic bending of the lithosphere in the front of the subduction zone encounters several difficulties. First of all, to derive the relevant solution, one should "cut" the subducting plate somewhere around the trench axis (or deeper) and fix some boundary conditions at the cut. These conditions are of course not known and were chosen to get the best fit between theoretical and actual relief. The parameters of the plate are also unknown and have been varied to obtain best fit between theoretical and natural data. These parameters for the elastic plate model are the lithospheric ("elastic") thickness H and the effective modulus of elasticity E. Different sets of E and H can produce identical theoretical profiles. The same is true for a more realistic elastico-plastic lithospheric model.

A different formulation of the problem was considered by *Tishchenko* [1985] and *Kerchman et al.* [1986]. Numerical modeling was used to study the initial stage of subduction along a fixed inclined "cut" in the lithosphere. This formulation is similar to the initial stages of experiments 3.6, 3.7 and 3.8 (Figures 3.7, 3.8 and 3.9). The results are similar as well. The model does not require arbitrary boundary conditions to be fixed. However, another difficulty arises due to the limitation of the theory which can be applied only to a small deformation corresponding to the earliest stages of subduction. Our experiments have shown that the initial subduction is essentially non steady state with relief being considerably different from that in mature subduction. As shown in Figure 3.7*b*, the maximum bending of the subducting plate at the beginning of subduction occurs in the upper part of the outer trench slope (see the arrow in Figures 3.7*b* and 3.7*c*), while practically no bending is observed in the lower part. Steady state shape of the bending which is established later on is characterized by shallower and narrower trench, with maximal curvature of the plate shifting toward the trench. Thus, the estimates for the lithospheric parameters σ_s, E and H derived by comparing the mentioned mathematical models with actual bathymetric profiles can hardly be regarded as reliable. Certain difficulties also arise when experimental modeling is used. A comparison of experiments 3.6, 3.7 and 3.8, Figures 3.7 to 3.9 yields a fairly reliable estimate for σ_s, as the plate shape in the subduction zone strongly depends on that parameter. As mentioned earlier, the best fit between the model and the generalized relief in the actual subduction zones is obtained in experiment 3.8 (Figure 3.9) for $\sigma_s^m \approx 2\tau_s^m = 30$ Pa. Adopting $H^0 = 60$ km as lithospheric thickness, we use criteria (2.1) to get a value of $\sigma_s^m = 2.6 \times 10^8$ Pa for the actual lithosphere. Similar values have been derived by other authors from gravity data [*Lobkovsky and Sorokhtin*, 1976; *Ushakov*, 1968].

The effective modulus of elasticity E for the plate has also been varied. However, no obvious and stable changes in the deformation of the lithospheric model have been detected. Such changes can occur only in the section which involves quasi-elastic plate bending in the outer rise area where small elastic strain plays an important part (elastic properties are not important in the zone of large plastic deformation under the overriding plate). Vertical displacements in that area are very small in the experiments, which makes it impossible to determine reliably the relationship between *E* and geometry of the outer rise. To estimate the effect of *E*, numerical modeling of the initial stage of subduction for an elastico-plastic plate was performed using the finite element method [*Tishchenko*, 1985]. The value of σ_s^0 was fixed to be that which followed from experimental modeling, while E^0 was varied in the range $E^0 = 10^{10} - 10^{11}$ Pa. The amplitude of plate convergence (subduction) was constant in all cases and was equal to 10 km. The results are shown in Figure 3.23. The influence of *E* on the "relief" indeed has proved to be small, and thus all the values used in the computation are acceptable (this conclusion probably remains valid also for mature subduction stages). Outside of that range, the amplitude of the outer rise is

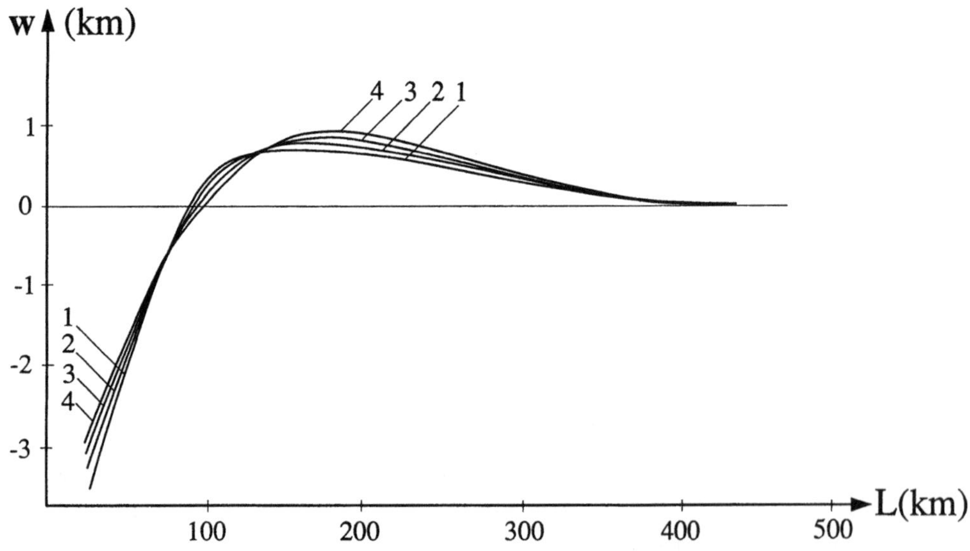

Figure 3.23. An elastico-plastic bending of the plate in front of the subduction zone, results of numerical modeling for various *E* values (from *Tishchenko* [1985])
(1) $E = 0.11 \times 10^{11}$ Pa; (2) $E = 0.23 \times 10^{11}$ Pa; (3) $E = 0.46 \times 10^{11}$ Pa; (4) $E = 0.92 \times 10^{11}$ Pa; (*w* is deflection of the plate surface from the isostatic equilibrium level; *L* is distance from the trench)

either too small or too large. The *E* value which was used in the experiments described in this work lies within the above range. The values for lithospheric parameters assumed at the beginning of this chapter $\sigma_s^o = 2.6 \times 10^8$ Pa and $E^o = 1.7 \times 10^{10}$ Pa thus follow from modeling results. At the same time they are consistent with the results of other workers [*Sorokhtin*, 1974; *Lobkovsky and Sorokhtin*, 1976; *Ushakov*, 1968].

It is interesting to note that the width of the outer rise is 200 to 400 km on average. Therefore the plate bending wavelength λ is about 600 to 900 km. On the other hand, the buckling instability for the lithosphere in the zone of intraplate deformation in the Indian Ocean has a wavelength that is smaller by three or four times. The difference is due to a different horizontal compression of the lithosphere. With small or no compression, λ reaches values of the order of 1000 km, while when the compression is close to σ_s, the λ value diminishes to a few hundred kilometers. We conducted experiments in which the subduction was blocked at some stage (this was done by local cooling of the model material in the contact zone between the plates). Since subduction became impossible, the further motion of piston 4 (Figure 2.1) led to a greater compression of the model which resulted in a decrease in λ and an increase in the amplitude of both the outer rise and the flexural sag seaward of it. Subsequent compression resulted in localization of deformation in this sag similar to that in experiments 3.3, 3.4 and 3.5, Figures 3.3, 3.5, and 3.6 (see also Figure 5.1).

3.9. Conclusions

The main conclusions of this chapter are the following:

1. Strong compression of the lithosphere produces flexural buckling with wavelengths of a few hundred kilometers. A localization of deformation occurs then in one of the flexural downwarps resulting in thickening of the lithosphere in that location, formation of the ridge and two major inclined, intersecting thrust zones crossing the whole plate thickness. The plate then fails along one of these planes initiating subduction. The inner trench slope which forms during this process has inhomogeneous step-like structure. Its lower part comprises a deformed block of oceanic crust and sediments. The block forms the slope break which exerts a damming effect for the arc-derived sediments resulting in the formation a deep sea terrace.

2. Modeling of the intraplate deformation in the northeastern Indian Ocean corroborates an earlier hypothesis concerning flexural buckling of the lithosphere and possible initiation

of subduction zone in that area.

3. If there is a pre existing inclined zone of weakness (fault zone) in the oceanic lithosphere sub-perpendicular to the axis of compression, subduction will be initiated along this fault. The formation of a subduction zone in such a situation requires lower compression than in the preceding case by a factor 2 or 3. The inner trench slope has a different "extensional" structure and forms due to normal faulting in the frontal part of the overriding plate.

4. In contrast to the overriding plate where the greatest deformation develops during the initiation of subduction, the subducting plate experiences large quasistationary deformations all the time during subduction. The plate undergoes bending in the trench-outer rise area which increases toward the trench. In the area of maximal bending (under the frontal segments of the overriding plate) intensive thrust deformation can develop along the zone dipping under the ocean and crossing the entire lithosphere.

5. The subducting plate exerts nonhydrostatic pressure on the overriding plate which causes positive free air gravity anomalies in the area of frontal arc amounting to a few hundred mGals.

6. Comparison of the relief in subduction zone at different parameters of the model lithosphere with the generalized relief in actual subduction zones provides the following estimate for the effective parameters of oceanic lithosphere of thickness $H = 60$ km: the shear yield limit $\tau_s = 1.3 \times 10^8$ Pa and the modulus of elasticity E of about a few times 10^{11} Pa.

CHAPTER 4

RELATIONSHIP BETWEEN SUBDUCTION REGIME AND BACK ARC DYNAMICS

4.1. Introduction

In all experiments presented in the previous chapter densities of the lithosphere ρ_l and the asthenosphere ρ_a are equal. Such a condition corresponds to reality only to a first approximation. In fact, ρ_l may be either less than or greater than ρ_a. The lithospheric density depends on many factors, the "thermal" age being the most important one [*Oxburgh and Parmentier*, 1977; *Oxburgh and Turcotte*, 1976; *England and Wortel*, 1980]. The greater the age of lithosphere (i.e., the colder and thicker it is), the higher its average density ρ_l. Estimates for $\Delta\rho$ ($\Delta\rho = \rho_l - \rho_a$) vary from very small values to those in excess of 0.1×10^3 kg/m^3 [*Yoshii*, 1973; *Toksoz et al.*, 1973; *Grow*, 1973; *Watts and Talwani*, 1975]. The average lithospheric and asthenospheric densities become equal when the age of the oceanic lithosphere reaches about 50 Ma [*Oxburgh and Parmentier*, 1977; *Molnar and Atwater*, 1978; *England and Wortel*, 1980; *Sacks*, 1983].

In this chapter we report the results of modeling when $\Delta\rho$ is varied. The experiments reveal that even a minor departure of ρ_l from ρ_a leads to considerable changes in the subduction regime and to interesting mechanic (geodynamic) effects. Some more subduction-related phenomena such as the opening of back arc basins, the relation of back arc dynamics to relief, the gravity field, etc. can be explained by the results of this next approximation in subduction modeling [*Shemenda*, 1985, 1986 and 1993].

4.2. Experimental Results

We varied the density ρ_l^m of the subducting plate within the limits 0.86×10^3 kg/m^3 to 1.05×10^3 kg/m^3 while density of the asthenosphere ρ_a^m was held constant, equal to 1×10^3 kg/m^3 as in previous runs. The early stages of the experiments proceeded similarly to the earlier trials. In all cases, compression of a continuous model lithosphere results in flexural buckling of the plate, localization of deformation in one of the sags, failure of the lithosphere along an inclined surface (fault) and finally, in the initiation of subduction. However the further process, i.e., subduction proper, depends largely on $\Delta\rho$. The results of a set of experiments conducted with different $\Delta\rho$ values ($\Delta\rho^m = 0$; $\Delta\rho^m < 0$; $\Delta\rho^m > 0$) are presented below. An inclined transverse cut dividing the lithospheric model into an overriding and a subducting plates was made beforehand in these experiments (Figure 4.1*a*) for simplicity.

Experiment 4.1. $\Delta\rho^m = 0$ ($\rho_l^m = \rho_a^m$). The subduction in this experiment is steady state except in the earliest stages (Figure 4.1). The relief produced by this process is also stable including the elevation of the frontal arc (see also Figure 4.2, dotted line). The subducted plate (Benioff zone) dips at a constant angle during subduction due to hydrostatic equilibrium with the surrounding mantle.

A decrease in lithospheric density ρ_l modifies the lithospheric deformation. The elevation of the frontal arc becomes higher and grows during the subduction. Also, the subducted plate starts to float up when it reaches a specific length. The lower the ρ_l value (i.e., the greater the $|\Delta\rho|$), the earlier the "Benioff zone" begins to rise and the higher the elevation of the frontal arc is. This is demonstrated in the next two experiments.

Experiment 4.2, $\Delta\rho^m = -0.14 \times 10^3$ kg/m^3. The buoyancy force is so large in this experiment that the subducted plate begins to rise almost immediately (Figure 4.3). The subduction (or more exactly the underthrusting) ceases after some time and all of the deformation is concentrated in the intraplate regions, mostly within the subducting plate (Figures 4.3*c* to 4.3*e*). The deformation then localizes in the sag behind the outer rise resulting in failure of the lithosphere and the formation of a new subduction zone (Figure 4.3*e*) (see also Section 5.2 and Figure 5.1). The value of $\Delta\rho^m$ is too large in this experiment, being $\Delta\rho^o = -0.46 \times 10^3$ kg/m^3 when converted to nature, and may only correspond to subduction of the continental lithosphere with thick light crust. The next

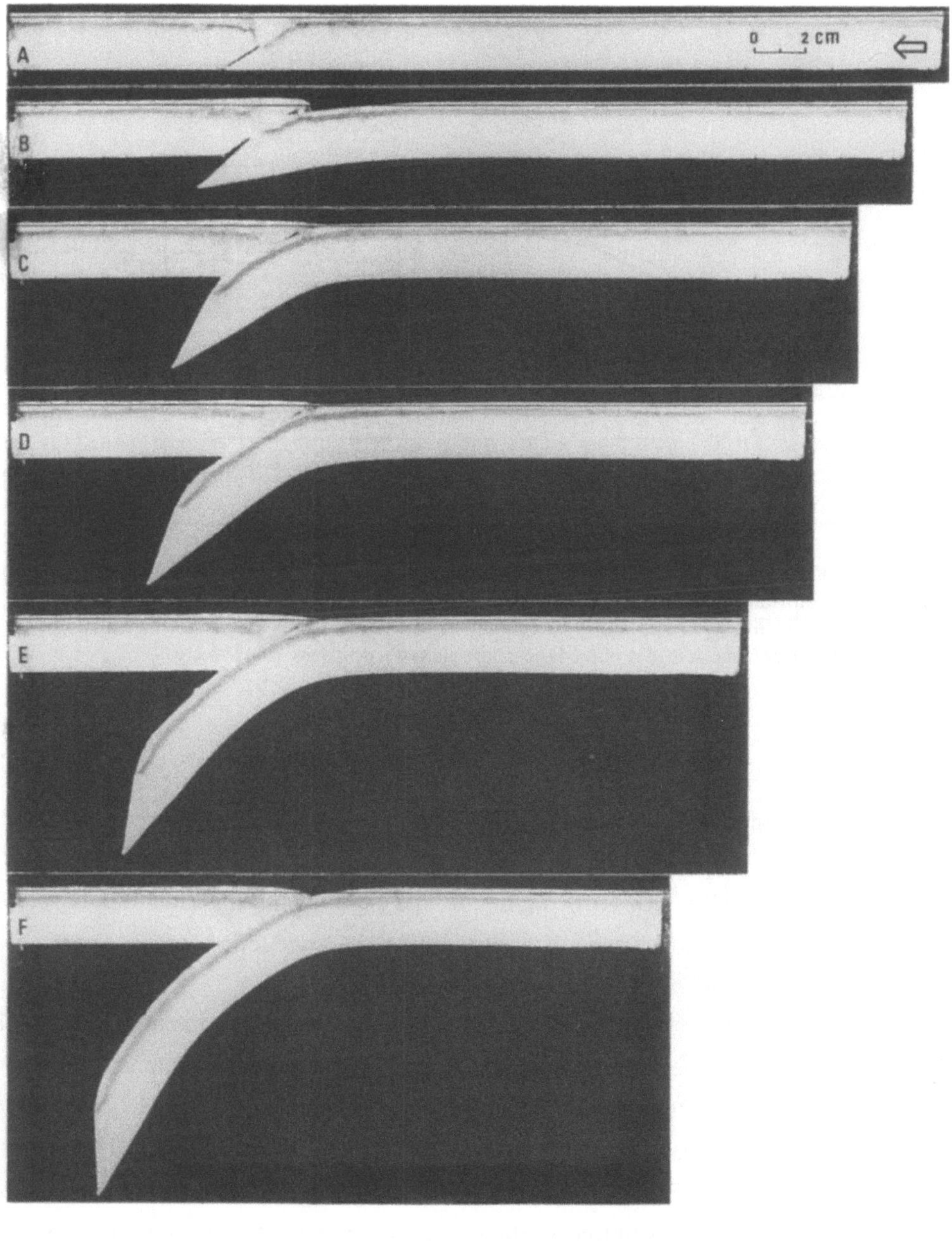

Figure 4.1. Experiment 4.1; subduction of a "normal" plate ($\Delta\rho^m = \rho_l^m - \rho_a^m = 0$).

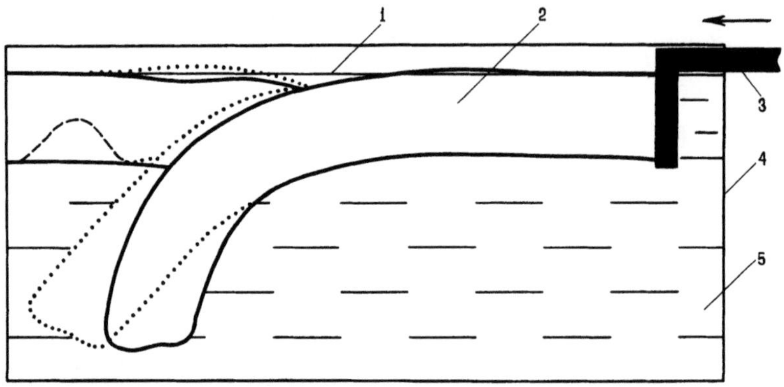

Figure 4.2. Configuration of the model lithosphere in subduction zone
1, isostatic equilibrium level of the "lithospheric" surface; 2, "lithosphere"; 3, piston; 4, tank; 5, "asthenosphere". The lithosphere as shown by continuous lines corresponds to $\rho_l > \rho_a$; the dotted line is for $\rho_l = \rho_a$; ρ_l and ρ_a are the densities of the lithosphere and asthenosphere, respectively.

experiment is conducted with $\Delta\rho^m$ value which is more realistic for the oceanic lithosphere.

Experiment 4.3, $\Delta\rho^m = -0.07 \times 10^3$ kg/m^3. In this experiment the subduction develops more easily (without intensive deformation out of the subduction zone), although a flattening of the "Benioff zone" again occurs (Figure 4.4).

We also conducted experiments with the same $\Delta\rho$ value (as in experiment 4.3), but with a lower yield limit σ_s. The result was similar to experiment 4.2. Two parameters $\Delta\rho$ and σ_s thus control the resultant pattern of the lithospheric deformation.

Experiment 4.4, $\rho_l > \rho_a$; $\Delta\rho^m = 0.03 \times 10^3$ kg/m^3 ($\Delta\rho^o \approx 0.1 \times 10^3$ kg/m^3). Subduction in this experiment is accompanied at an early stage by elevation (formation) of the frontal arc (Figure 4.5*b*) similar to the previous experiments. The amplitude of this elevation is, however, considerably smaller. Unlike experiments 4.1 to 4.3 the frontal arc

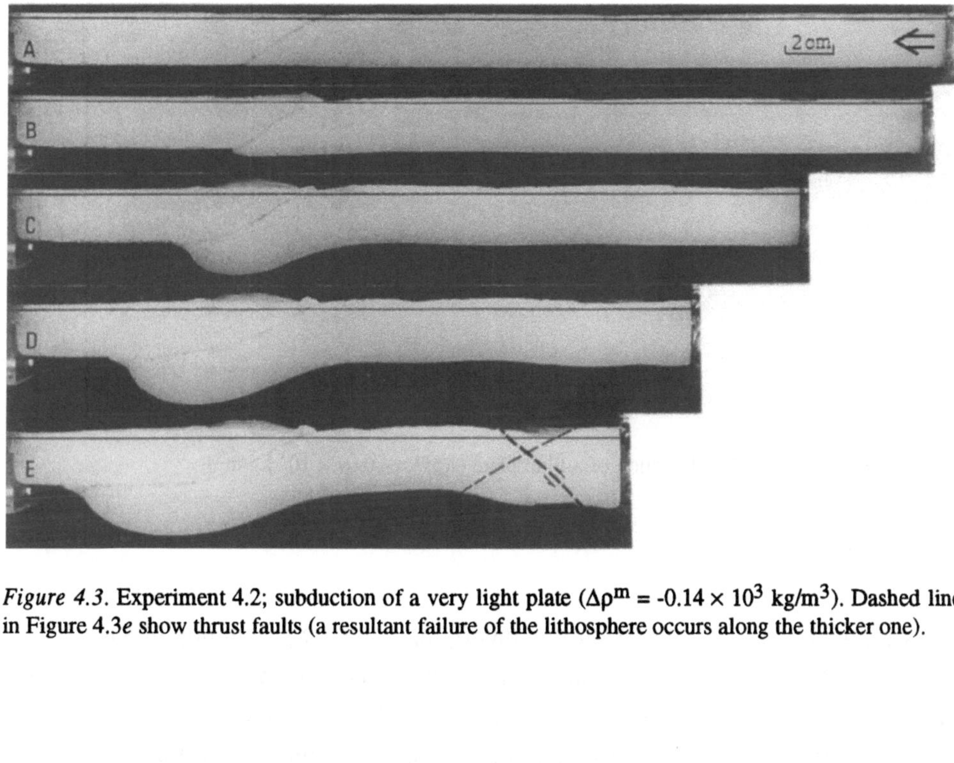

Figure 4.3. Experiment 4.2; subduction of a very light plate ($\Delta\rho^m = -0.14 \times 10^3$ kg/m^3). Dashed lines in Figure 4.3*e* show thrust faults (a resultant failure of the lithosphere occurs along the thicker one).

starts to subside during further subduction sinking below the level of the isostatic equilibrium of the plate (see Figures 4.5*c*, 4.5*f* and Figure 4.2, the solid line). The dip angle of the subducted plate progressively increases. When the length of the subducted segment of the oceanic lithosphere reaches a certain value, it breaks directly beneath the interplate contact zone (Figure 4.5*f*, dashed line). Both the overthrusting wedge and the trench rise after that and then the whole cycle is repeated again.

We determined the nonhydrostatic horizontal stresses σ_h in the lithosphere in the above experiments using the procedure described in Section 2.4. Numerous test experiments were conducted for that purpose with progressive thickening of the notch at the base of the overriding plate. In these experiments the overriding plate was made of a material with

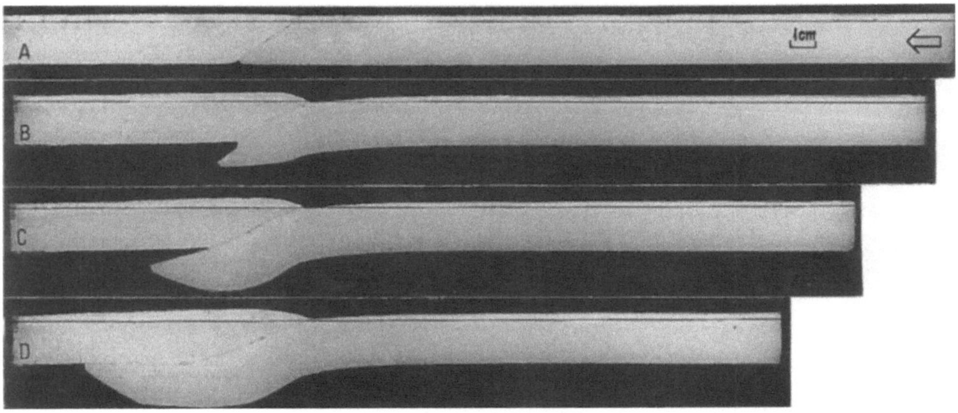

Figure 4.4. Experiment 4.3; subduction of a light plate ($\Delta\rho^m = -0.07 \times 10^3$ kg/m^3).

density ρ_o which is equal to that for the underlying mantle ρ_a. Thinning of such a plate from below does not disturb its initial stress state because the removed part of the lithosphere is replaced by a fluid of the same density. If the notch were made from above or if $\rho_o \neq \rho_a$, then the procedure for determining σ_h used would be invalid. We obtained the following results. For $\rho_l = \rho_a$ (experiment 4.1, Figure 4.1) $\sigma_h = (0.3 - 0.4)\sigma_s$ (σ_s is the lithospheric yield limit under normal load), except at the earlier stages of the process when σ_h is somewhat higher. Decreasing ρ_l, increases σ_h. Compression close to the lithospheric yield limit is necessary to initiate subduction (underthrusting) under the conditions of experiment 4.2 (Figure 4.3), that is, when $\Delta\rho^m = -0.14 \times 10^3$ kg/m^3. The compression increases in this experiment while subduction develops and reaches σ_s. After that, the specimen fails in a new location, with the subduction zone jumping to that place. In experiment 4.3 with $\Delta\rho^m = -0.07 \times 10^3$ kg/m^3, σ_h is somewhat lower.

Stress σ_h in experiments 4.2 and 4.3 is largely determined by the buoyancy force; the greater the $|\Delta\rho|$ value, the higher σ_h. Therefore the ratio σ_h/σ_s increases for decreasing σ_s

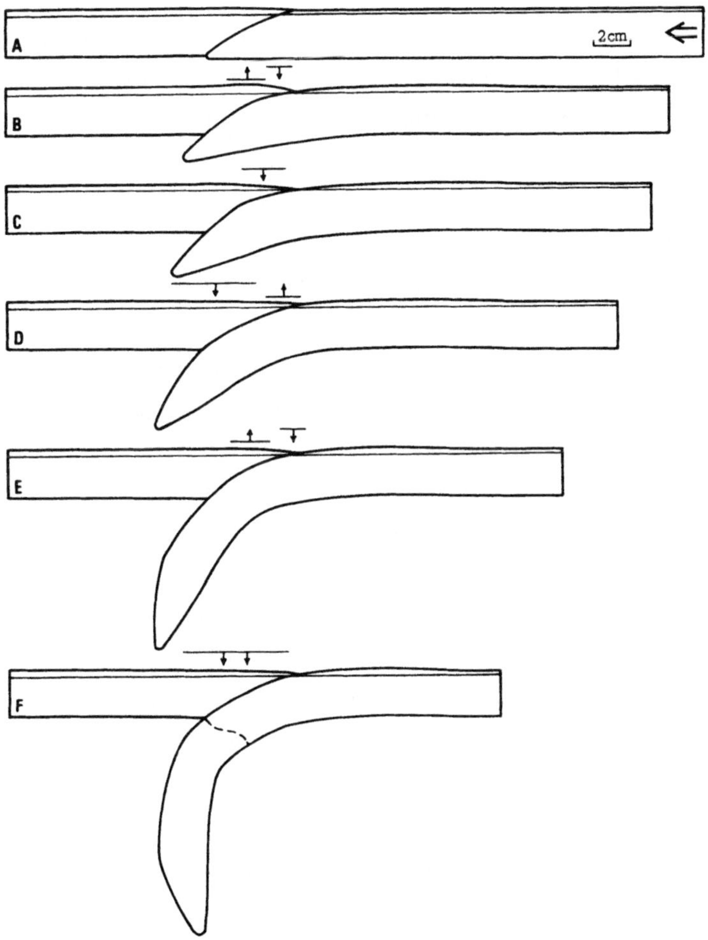

Figure 4.5. Experiment 4.4; subduction of a dense plate, $\Delta\rho^m = 0.03 \times 10^3$ kg/m^3 (drawings of photos). The dashed line in Figure 4.5*f* shows the place where the subducted plate breaks after its length reaches a certain value. Arrows show directions for the vertical motion of the relevant segments of the lithosphere surface.

with $\Delta\rho$ = constant. Also, lower strength of the lithosphere makes it easier for the buoyancy force to bend the subducted plate causing its movement toward the surface. That is why decreasing σ_s with the same $\Delta\rho$ value as in experiment 4.3 makes the deformation of the lithosphere similar to experiment 4.2 in which the buoyancy force is larger.

A more complex pattern of σ_h variation is in experiment 4.4 (Figure 4.5) where $\rho_l > \rho_a$. The lithosphere experiences a small compression of $\sigma_h \approx 0.1\sigma_s$ during the earlier stages of the experiment when the frontal arc is slightly uplifted. The compression then decreases when both the length of the subducted plate increases and subsidence develops in the frontal arc. At some stage of the subduction, compression is replaced by tension which progressively increases during further subduction. This continues until the subducted plate hanging into the mantle fails under its own weight. Thereafter tension is again replaced by compression and the whole cycle repeats itself from the beginning. The maximum tensile stresses reaches $\sigma_h = (0.1 - 0.2)\sigma_s$, which when converted to nature is about $\sigma_h = 0.4 \times 10^8$ Pa.

In the next experiment, the same parameters were used as in experiment 4.4. The difference is that the overriding plate is thinned sufficiently to be failed under the tensile stress σ_h that occurs during subduction.

Experiment 4.5, $\rho_l^m > \rho_a^m$; $\Delta\rho^m = 0.03 \times 10^3$ kg/m^3. The earlier stages of this experiment (Figure 4.6) proceed as in experiment 4.4. However, after the length of the subducted segment of the lithosphere reaches a certain value, the overriding plate breaks apart in the thinned zone (Figure 4.6*c*). The broken "arc plate" starts to move toward the subducting plate sliding along its surface and causing back arc spreading (the piston pushing the subducted plate to the left keeps moving at a constant rate). This process lasts until the oceanic plate rests against the bottom of the bath (Figure 4.6*e*).

4.3. Hydrostatic suction of the plates

The results of experiments 4.1 to 4.3 seem fairly understandable while the results of experiments 4.4 and 4.5 were somewhat unexpected. The main question that arises is why does the gravitational force pulling the subducted plate downward causes extension and failure in the overriding plate instead of separation of the plates in the subduction zone. One can suppose this effect to be caused by molecular adhesion of the plates or by other factors which are not significant in nature. To gain insight into this question and to understand the physical mechanism responsible for the result obtained we conducted

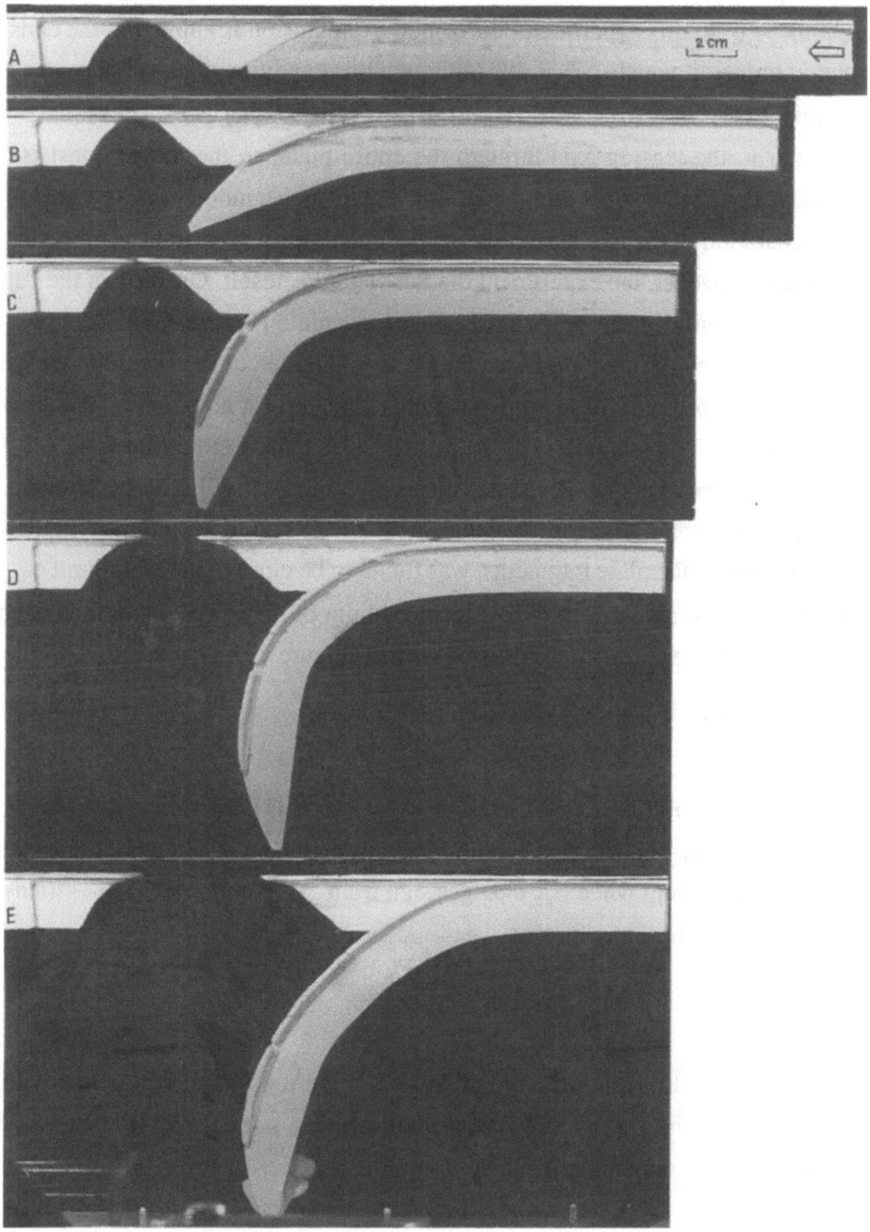

Figure 4.6. Experiment 4.5; subduction of a dense plate ($\Delta\rho^m = 0.03 \times 10^3$ kg/m^3). This experiment differs from the previous one in that there is a pre existing thinning of the overriding plate from below (density of the overriding plate ρ_o^m is equal to the density of the underlying mantle, $\rho_o^m = 10^3$ kg/m^3).

another, simpler experiment (for details see *Malkin and Shemenda* [1991]).

Experiment 4.6. In this experiment, horizontal axial tension is applied to the ends of the lithospheric model in which $\rho_l^m = \rho_a^m$ (although this condition does not play an essential role in this case). A cross vertical cut BD (Figure 4.7*a*) was previously made, perpendicular to the tension axis through the entire plate, so the model consists of two separate halves pressed against each other. Tension of such a model was expected to result in simple separation of the two halves and in the underlying asthenosphere filling the space between them. However, the experiment did not yield this result. Opening of the cut under tension occurred only in the uppermost part of the plate (Figure 4.7*a*). Below, plastic deformation of the model developed, leading to the formation of an upwelling wedge ABC (Figure 4.7*a*). The presence of the pre existing cut had almost no effect on this lower part of the lithosphere. Complete failure of the plate occurred along either slip zone AB, BC, or both at once. Deformation then developed either asymmetrically (Figures 4.7*b* and 4.7*c*) or symmetrically (Figure 4.7*d*). The tensional stress σ_h needed to accomplish the described process and separate the plate fragments, was found to be close to the yield limit σ_s.

Let us consider this result from the mechanical point of view. The condition for opening the cut in the plate in experiment 4.6 can be written (Figure 4.8) as

$$\sigma_{xx} = 0, \tag{4.1}$$

where σ_{xx} is the normal horizontal stress and the x-axis is perpendicular to the cut (i.e., parallel to the tension vector). This condition means that the walls of the cut separate from each other at the moment when σ_{xx} changes sign from "-" (compression) to "+"(tension). Condition (4.1) can also be written as

$$\sigma'_{xx} - \rho_l g z = 0,$$

where $\rho_l g z$ is the hydrostatic pressure at depth z (Figure 4.8) and σ'_{xx} is nonhydrostatic normal stress along the x-axis. Stress σ'_{xx} obviously can not exceed σ_s. Therefore (4.1) can hold, down to the maximum depth

$$z_m = \sigma_s / \rho_l g \tag{4.2}$$

(hydrostatic pressure does not allow the cut to open at greater depths). Substituting the

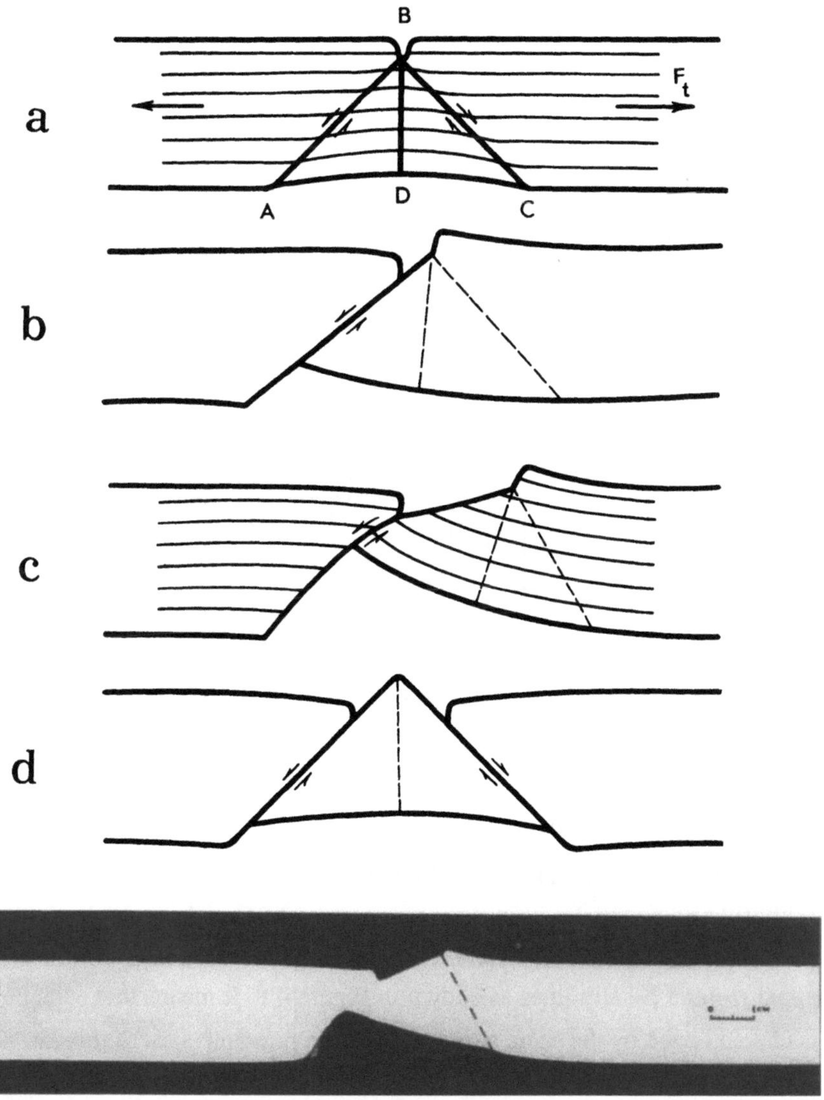

Figure 4.7. Experiment 4.6; extension of the lithosphere with a vertical cut BD (Figure 4.7*a*) through the whole plate transverse to the tension axis
(*a*) to (*c*) A variant of asymmetric deformation (in cross-section) when failure occurs along a surface, either AB or BC (in the case shown, along AB). (*d*) Initial stage of the deformation when the failure occurs along both surfaces, AB and BC. (*e*) A photo of the deformed model corresponding to stage (*c*) .

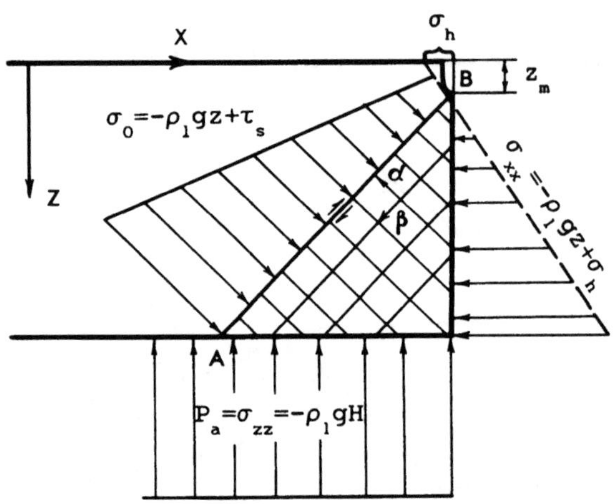

Figure 4.8. Stress in the plate and slip lines in the plastic zone for experiment 4.6.

model parameters $\sigma_s^m = 30$ Pa and $\rho_l^m = 10^3$ kg/m^3, we get $z_m \approx 3$ mm. This is close to what is observed in the experiment 4.6 (Figure 4.7*e*).

The lithospheric material within the rising wedge ABC undergoes plastic deformation before the complete failure of the plate. The stress state in this area is uniform with straight orthogonal α– and β– slip lines as shown in Figure 4.8. It means that each point of this zone is characterized by the same stress tensor. The principle axes of this tensor coincide with the *x*- and *z*-axes in Figure 4.8. The Mises-Hencky yield criterion which is held in the plastic zone can be written for the approximation of the plane strain state as (see Section 3.7)

$$(\sigma_{xx} - \sigma_{zz}) = \sigma_s,$$

where σ_{xx} and σ_{zz} are appropriate principal stresses. At boundary AC (Figures 4.7*a* and

4.8) $\sigma_{zz} = -\rho_l gH$. Therefore σ_{xx} at this boundary is

$$\sigma_{xx} = -\rho_l gH + \sigma_s.$$

Nonhydrostatic stress σ'_{xx} throughout the plastic zone, including the boundary lines AB and BC, is $\sigma'_{xx} = \sigma_s$. Hence, the net force pulling the plate fragments apart is $F = \sigma_s(H - z_m)$ and the tensile stress σ_h far from the cut is $\sigma_h = F/H = \sigma_s(1 - z_m/H)$. Considering that z_m/H is small compared to unity, σ_h will approximate σ_s, as confirmed by experiment 4.6.

Thus, the plate fragments behave as if sucked together due to hydrostatic forces. The fragments can slide easily relative to one another, needing low shear stresses to overcome only the friction. On the other hand the hydrostatic pressure holding the plates together must be overcome to separate them. Plate strength proves to be insufficient to withstand such large tectonic stresses which leads to the plastic deformation and failure of the lithosphere.

Similar experiments have been conducted in which the model lithosphere was slightly weakened (thinned at the base by 20% - 30%) through the entire width at some distance outside of the cut perpendicular to the tension. It is in this weakened zone that plastic deformation and failure occur under tension in spite of the presence of the cut. The physical similarity of simple experiments presented in this section to experiments 4.4 and 4.5 (Figures 4.5 and 4.6) is that in both cases, the plates in contact along the interplate surface (vertical in Figure 4.7 and inclined in Figures 4.5 and 4.6) are under tension which would tend to separate them. In experiments 4.4 and 4.5 (Figures 4.5 and 4.6) tension occurs owing to the tendency of the heavy subducting lithosphere to move apart from the overriding plate. We analyze this case in more detail below.

4.4. Stress in the Overriding Plate Due to Subduction of an Old (Dense) Lithosphere

For simplicity we will neglect the rigidity of the overriding wedge of the arc plate. The full contact pressure P_n between the plates in the subduction zone at each point on AB (Figure 4.9*a*) can then be considered to be entirely determined by the weight of the overriding lithospheric material. To conform to this condition more strictly, special experiments have been conducted with vertical cross cuts having been previously made in

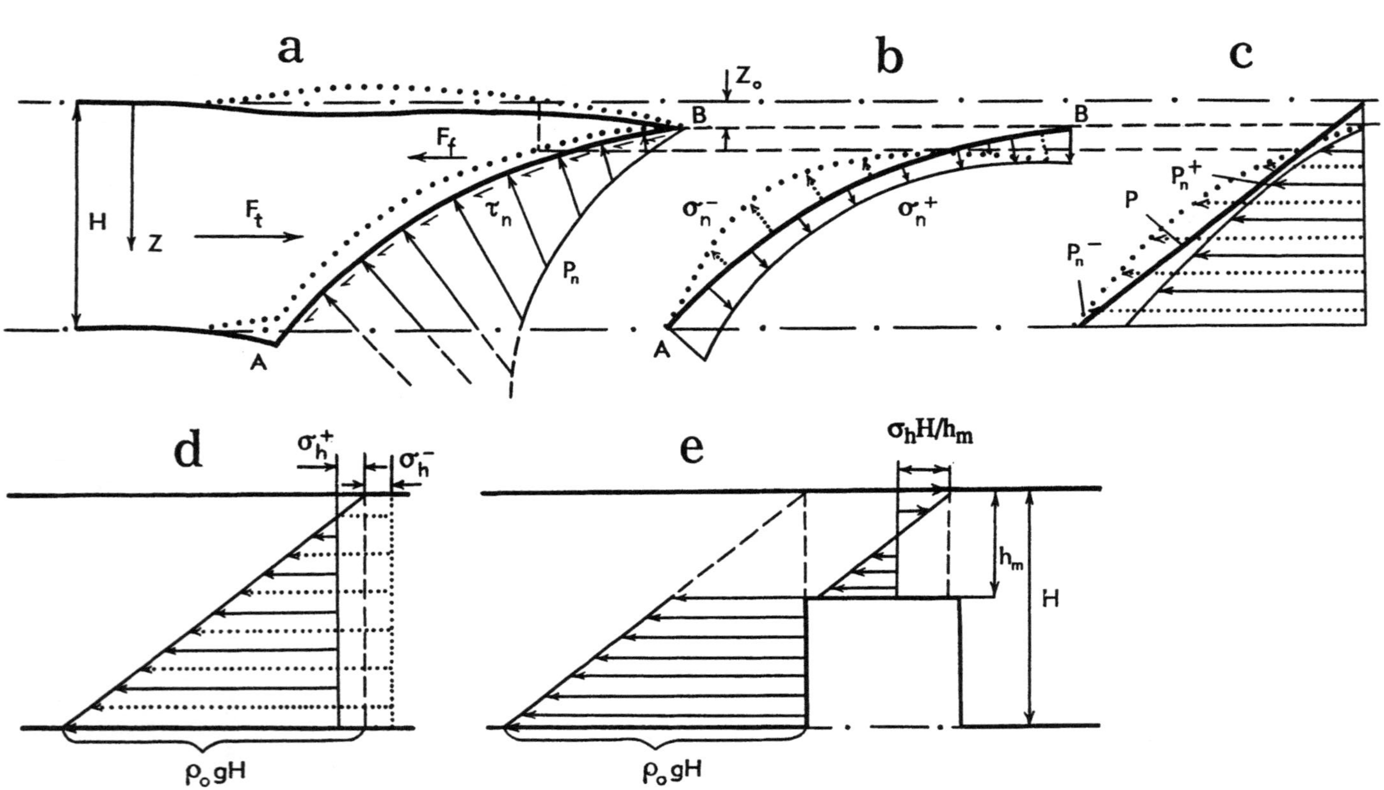

Figure 4.9. Stress in the overriding plate.
Continuous lines and superscript "+" correspond to $\rho_l > \rho_a$; dotted lines and superscript "-" are for $\rho_l = \rho_a$

(*a*) The form of the overriding plate and deflection of its surface from isostatic equilibrium level (dot-dash line) during subduction, P_n, τ_n are normal and tangential stresses acting on the overriding plate in the interplate zone AB. (*b*) Envelope of nonhydrostatic pressure σ_n in the interplate zone at each horizontal level. (*c*) Net contact pressure P_n on each horizontal level., P is the hydrostatic pressure $P = -\rho_0 gz$. (*d*) Envelopes for horizontal normal stresses in the overriding plate far enough from AB. (*e*) Concentration of tensional stresses in a thinned plate.

the overriding plate. Similar to Section 3.7 we will use further a nonhydrostatic pressure σ_n between the plates $\sigma_n = P_n - P$, where $P = -\rho_o g z$ and ρ_o is the density of the overriding plate ($\rho_o = \rho_a$). The σ_n value can be roughly inferred from the deflection w of the overriding wedge surface from the isostatic equilibrium level (see Figures 4.2 and 4.9*a*), $\sigma_n = \rho_o g w$ (before the experiment the surface coincides with the equilibrium level). Figure 4.9*b* shows σ_n along AB for mature subduction when $\rho_l > \rho_a$ (σ_n^+) and for $\rho_l = \rho_a$ (σ_n^-), as it follows from the position of the overriding wedge surface with respect to the equilibrium level in both cases (Figure 4.9*a*). A still higher position of the σ_n curve with respect to σ_n^- corresponds to experiments 4.2 and 4.3, where $\rho_l < \rho_a$.

The net lateral force exerted on the overriding plate from the interplate boundary AB can be written, neglecting friction τ_n, (Figures 4.9*a* to 4.9*c*) as

$$F = \int_{z_o}^{H} P_n(z)dz = \int_{z_o}^{H} \sigma_n(z)dz + \int_{z_o}^{H} Pdz, \tag{4.3}$$

where z_o is the trench depth with respect to the hydrostatic equilibrium level of the overriding plate. Taking into account the fact that the tangential stresses at the plate bottom are negligible due to low viscosity of the asthenosphere, and that the deflection (tilt) of the overriding wedge is small, one can consider the compressional force F to be constant along the overriding plate. Beginning at some distance from the boundary of disturbance AB (Figure 4.9), normal total horizontal stress σ_{xx} in the overriding plate can be written as $\sigma_{xx} = \sigma_h + P$, where σ_h is the nonhydrostatic part of σ_{xx} (see Figure 4.9*d*). Therefore, F can be also written as

$$F = \int_{0}^{H} \sigma dz = \sigma_h H + \int_{0}^{H} Pdz. \tag{4.4}$$

Combining (4.3) and (4.4), we get $\sigma_h = 1/H \int_{z_o}^{H} \sigma_n dz + P_o$, where

$$P_o = \rho_o g z_o^2 / 2H. \tag{4.5}$$

The z_o value in the model (z_o^m) is 1-2 mm. Therefore, P_o^m is approximately $P_o^m \approx 0.5$ Pa $= 2 \times 10^{-2}\sigma_s$. The amplitude of $|\sigma_n|$ as derived from the maximum deflection w_m of the overriding wedge surface in the experiments (Figure 4.9*a*), is between $\sigma_s/2$ and σ_s (these values are consistent with that obtained in Section 3.7 from the consideration of the plastic deformation in the overriding plate: $\sigma_n \approx \sigma_s/2$). Hence, $\sigma_h^- = 1/H \int_{z_0}^{H} \sigma_n^- dz + P_0$ must be negative (see Figure 4.9*b*), whereas $\sigma_h^+ = 1/H \int_{z_0}^{H} \sigma_n^+ dz + P_0$ is positive. In other words, at $\rho_l \leq \rho_a$, the overriding plate is under nonhydrostatic compression and at $\rho_l > \rho_a$, the lithosphere is under tension [*Shemenda*, 1985]. Thus, even though the interacting plates in a subduction zone are always pressed against each other ($P_n < 0$), tension (tensile nonhydrostatic stresses σ_h^+) can occur in the overriding plate when the subducting lithosphere has a comparatively high density. Subduction of a dense lithosphere decreases the horizontal compression of the overriding plate by decreasing the contact pressure P_n. The vertical compression remains the same, being equal to the hydrostatic pressure. This leads to horizontal extension of the overriding plate. Tensile stresses concentrate in the thinned zone of the plate (Figure 4.9*e*) resulting in failure of the lithosphere at that location.

The mechanism responsible for the hydrostatic suction of the plates is similar to adhesion of an ordinary suction-cup to a smooth wall. The adhesion is caused by the confining atmospheric pressure and a lowering of pressure between the suction-cup and the wall surface. Some force is needed to remove the suction-cup from the wall, whereas moving the cup along the wall (if it is smooth, or moreover, if lubricated) does not require any effort. In our experiments 4.4 and 4.5, the role of the atmospheric pressure is played by the hydrostatic pressure in the lithosphere. If the strength of the wall (in the example with suction-cup) is assumed to be so small such that the removal of the suction-cup results in the wall being deformed, the indicated analogy would be even more exact.

Let us turn to the experimental results. The friction τ_n on the boundary AB (Figure 4.9*a*) was supposed to be absent. In fact (at $\tau_n \neq 0$), there is a horizontal component F_f of the friction force which counteracts the force F_t (Figure 4.9*a*) caused by the hydrostatic suction of the plates. Hence, extension of the overriding plate is possible only when $F_t > F_f$. Note that the horizontal force $\sigma_s h_m$ in the overriding plate as found in the experiments is a resultant of F_t and F_f.

To summarize, extension in the overriding plate during subduction of the dense (old) lithosphere (experiments 4.4 and 4.5) is not a consequence of the molecular adhesion of the plates. It is caused by hydrostatic suction due to the presence under the lithosphere of an effectively fluid substratum and hydrostatic stresses in the plate that are large, compared to the yield limit σ_s. If the lithosphere had much greater strength, this effect would not "work". There is no doubt (and it is easy to verify) that if plates under conditions in experiment 4.6 (Figure 4.7), were stronger, for example being made of metal, they would simply be separated from one another under tension. Indeed, the depth z_m of a cut opening increases according to (4.2) while σ_s grows, and can reach the plate thickness H when $\sigma_s \geq \rho_l g H$.

4.5. Other Mechanisms for Tension in the Overriding Plate

Above we considered only one situation in which back arc tension occurs due to the subduction of lithosphere which is denser than the mantle. However, due to the suction between the plates, back arc extension must also occur whenever plates in a subduction zone tend to move apart under any other force. For example tension can be produced by a horizontal mantle flow acting on the upper surface of subducted plate (a Benioff zone) [*Nelson and Temple*, 1972; *Packham and Falvey*, 1971; *Uyeda*, 1984; *Shemenda*, 1986]. Tension can also occur when the overriding plate moves in the same direction as the subducting lithosphere, but at a lower rate. Owing to the "sail" effect of the subducted plate in the asthenosphere, the subduction zone behaves as if it were anchored to the mantle [*Tullis*, 1972]. We also tried to model such situations. Tension in the overriding plate in experiments occurred even for $\rho_l < \rho_a$. Its magnitude was strongly dependent on the manner in which the "mantle" flow was generated as well as on the presence and width of the space between the "Benioff zone" and the bath walls. Normally the tensile stresses were invariably greater than for simple subduction of a dense plate.

Still greater tensional stresses reaching values of up to $\sigma_h \approx 0.4\sigma_s$ occur during pulling out of the subducted plate back from under the overthrusting lithosphere. Such an "aduction" process is modeled in the next experiment.

Experiment 4.7. This experiment represents a continuation of experiment 4.1 (Figure 4.1) after the direction of piston motion was reversed. As seen from Figure 4.10, this change did not produce separation of the plates in the subduction zone owing to the hydrostatic suction between them.

4.6. Influence of Crustal Thickness and Density of the Overriding Plate on Its Stress State

To imitate subduction under the continental lithosphere we performed experiments with density of the overriding plate ρ_o lower than density of the mantle ρ_a at various ρ_l values.

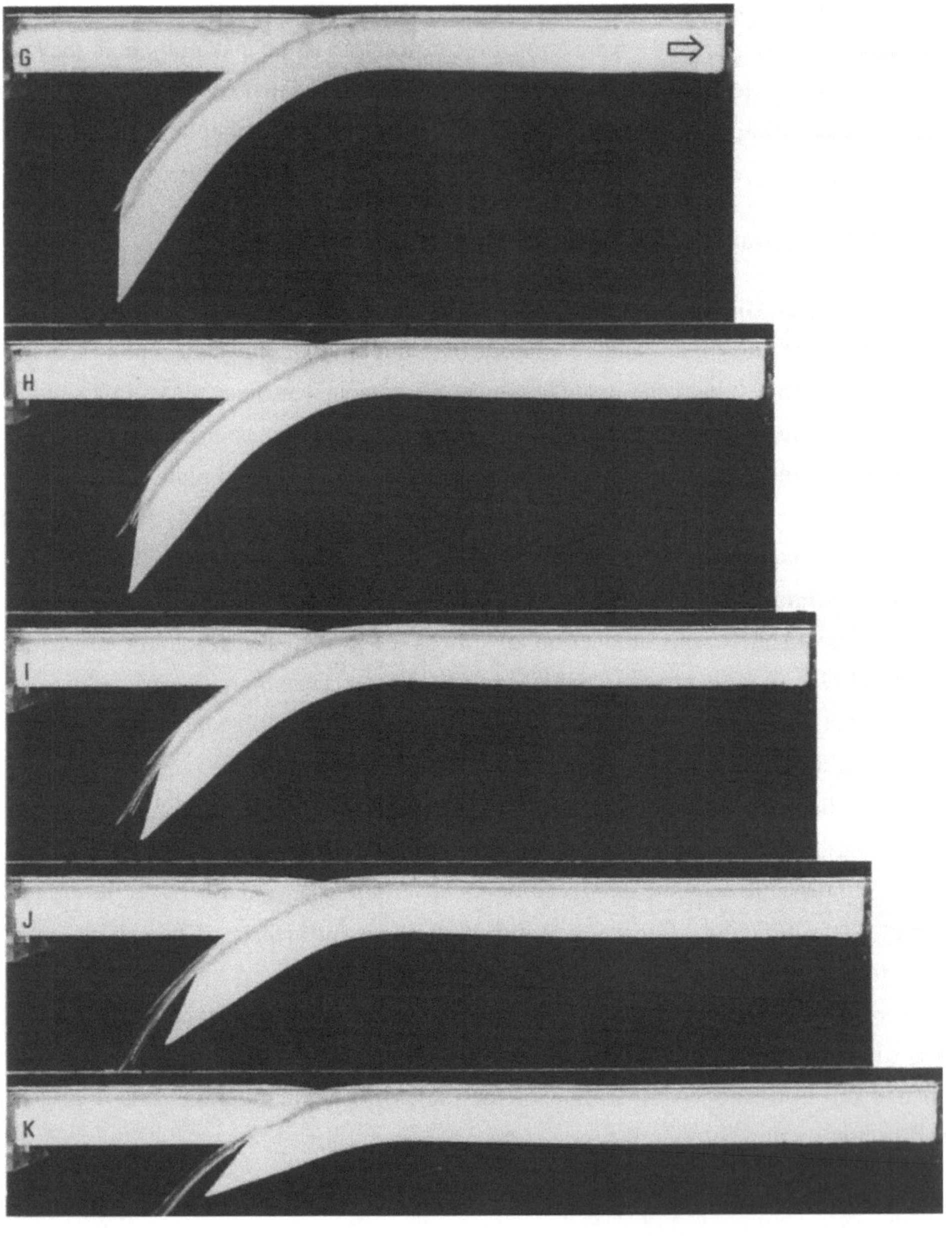

Figure 4.10. Experiment 4.7, reverse movement of the subducted plate (aduction), $\rho_l^m = \rho_a^m$. This experiment is a continuation of experiment 4.1 (Figure 4.1) with the direction of piston motion reversed.

The experiments showed that tension in the lighter overriding plate occurring during subduction of a dense lithosphere is more intense and back arc rifting develops more easily. Unfortunately, it was very difficult to make quantitative estimate of the relationship between the tensional force and ρ_0 because of the low accuracy of the experiments. However, to a first approximation, such a relationship can be inferred theoretically [*Shemenda*, 1985]. The excess pressure P_0 in (4.5) which enters the expression for normal horizontal stress σ_h in the overriding plate is always positive (i.e., tension). This parameter may be looked upon as a measure of the overriding plate's tendency to flow laterally towards the trench. Its value is controlled by the trench depth z_0 with respect to the isostatic equilibrium level of the overriding plate surface. It is clear that the thicker the crust of this plate, the greater z_0 would be, and consequently, the greater the spreading effect. The effect can be easily assessed. Let us consider for instance, a simple model shown in Figure 4.11 involving a two-layer overriding lithosphere and a one-layer subducting oceanic plate. Let ρ_l be the density of both the subducting lithosphere and the mantle part of the overriding plate, ρ_c the crustal density, H and H_c the thicknesses of the entire overriding plate and its crustal layer, respectively, and z_0 the trench depth with respect to the equilibrium level of the overriding plate. The horizontal spreading force for the overriding plate caused by internal hydrostatic pressure can then be written

$$F_1 = \int_0^{H_c} \rho_c g z dz + \int_{H_c}^{H} \rho_l g(z - \zeta) dz. \tag{4.6}$$

This force is balanced by hydrostatic and tectonic (nonhydrostatic) forces acting from the subducting plate as well as by the strength of the lithosphere (by elastic stresses). The hydrostatic force is equal to

$$F_2 = \int_{z_0}^{H_c} \rho_l g(z - \zeta) dz + \int_{H_c}^{H} \rho_l g(z - \zeta) dz, \tag{4.7}$$

where ζ is the depth of the surface of the undisturbed subducting plate relative to that of the overriding lithosphere (Figure 4.11),

$$\zeta = H_c(1 - \rho_c/\rho_l). \tag{4.8}$$

The difference between F_1 and F_2 gives an excess force ΔF of free spreading for the overriding plate. The relevant spreading stress would be $P_o = \Delta F/H$. Considering (4.6), (4.7), and (4.8), we get

$$P_o = g/2H\{\rho_c H_c^2 - \rho_l(H_c - z_o)[z_o - H_c(1 - 2\rho_c/\rho_l)]\}. \tag{4.9}$$

Setting $H_c = 0$ here, we get (4.5), as expected. Putting $H = 6.5 \times 10^4$ m; $H_c = 3.5 \times 10^4$ m; $\rho_c = 2.7 \times 10^3$ kg/m^3; $\rho_l = 3.3 \times 10^3$ kg/m^3 and $z_o = 10^4$ m into (4.9), we get

$$P_o \approx 5 \times 10^7 \text{ Pa} \approx 0.2\sigma_s. \tag{4.10}$$

The value $z_o = 3 \times 10^3$ m, the typical trench depth relative to the surrounding sea floor, should be adopted when the crustal layer is absent in the overriding plate, that is, when $H_c = 0$. In this case which corresponds to our experiments, $P_o = 2.3 \times 10^6$ Pa $\approx 10^{-2}\sigma_s$. This is considerably lower then for actual situation (4.10).

One can investigate other more sophisticated models involving the solution for the relevant problem of elasticity theory which would result in somewhat different numerical estimates. However in principle the result would remain the same.

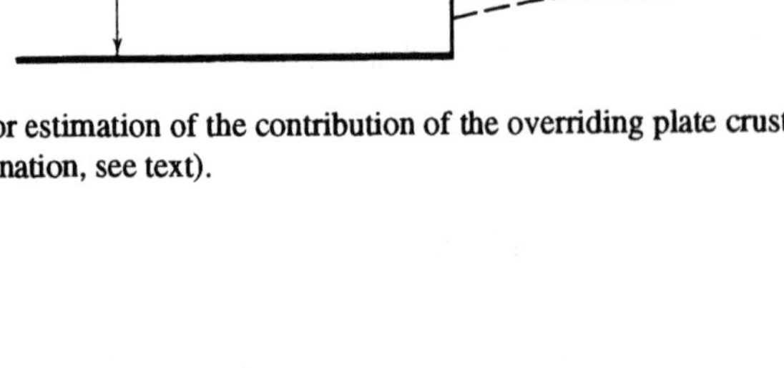

Figure 4.11. Scheme for estimation of the contribution of the overriding plate crust to back arc extension (for an additional explanation, see text).

Lower tectonic tensile stress σ_h is thus needed to break apart a lighter plate. Therefore, conditions for back arc spreading are more favourable when subduction develops beneath either the continental lithosphere or an island arc with a thick crust. On the other hand, greater tectonic stresses are required to break off the overriding plate in the intraoceanic subduction zones which are in early stages of evolution and consequently have a thinner crust [*Shemenda*, 1985; *Froidevaux et al.*, 1988].

It should be noted that the presence of a thick crust by itself is not sufficient to result in the formation of an active back arc basin. The action of tectonic tensional stress σ_h is necessary. The environment in the Central Andes corroborates this conclusion. The extension in the Altiplano seems to be a result of the effect of the lateral spreading of a thick crust [*Froidevaux and Isacks*, 1984]. This feature exists on a background of general lithospheric compression. If the compression is replaced by tension in the future, the Altiplano could develop into a back arc spreading basin.

4.7. Geodynamic Interpretation of Modeling Results

The modeling shown that back arc stress regime is largely dependent on two important factors which are the density difference $\Delta\rho$ and the relative motion of the lithospheric plates in the subduction zone and underlying mantle. Parameters such as the "thermal" age of the subducting plate, and density of the mantle in the subduction zone control the first factor. Increases in the age and/or decreases in ρ_a correspond to a transition from compression to tension in the overriding plate. The second factor operates as follows. The motion of the overriding plate in the mantle (hot spot) frame away from the subduction zone and/or the motion of the mantle substratum in the direction opposite to that of the subducting plate produce tension in the overriding plate. The opposite motions of the lithosphere and the underlying mantle tend to compress the lithosphere. All listed factors can be combined resulting in either amplification or reduction of the compression (tension).

It should be noted that the listed mechanisms have already been considered by different authors (mostly based on geological and geophysical data) as governing back arc dynamics. For example, a relationship established between the age of subducting plate and the mode of back arc dynamics had led to the conclusion that gravitational sinking of the dense subducting plate into the mantle can produce extension in the back arc area [*Molnar and Atwater*, 1978; *Uyeda and Kanamori*, 1979; *Dewey*,1980]. However, the answer has not been found as to why the sinking subducting lithosphere is not detached from the overriding plate, but pulls the island arc (arc plate) oceanward. Asthenospheric currents were considered as a possible mechanism governing the back arc regime in [*Nelson and Temple*, 1972; *Packham and Falvey*, 1971; *Uyeda*, 1984; *Shemenda*, 1986]. The hypothesis of the subducted plate "anchoring" was first suggested by *Tullis* [1972]. An anchoring results in back arc tension when the overriding plate moves away from the subduction zone in the mantle reference frame.

The experiments reported in this work provide a deeper understanding of the operation of all these mechanisms in delineating the effect of hydrostatic suction of the plates. It is important to note that the concept of "suction" has already been applied to explanation of back arc dynamics and overall of the driving mechanism of plate motion [*Elsasser*, 1971; *Forsyth and Uyeda*, 1975], however, its physical meaning has not been previously defined.

Other models are also suggested for the explanation of back arc spreading. *Karig* [1971] proposed that back arc spreading is caused by a diapir coming to the surface under a back arc region. *McKenzie* [1969], *Hsui and Toksoz* [1979] related back arc tension to a secondary (induced) convection above the sinking plate. Simple estimates [*Artushkov*, 1979] show that these mechanisms can hardly be the cause of back arc tension. At the same time induced convection can play an essential role in the weakening of the back arc lithosphere facilitating its breaking under tension (see below). Finally, back arc spreading can have a purely kinematic origin, that is, can be caused by relative motions of the plates and microplates and by unevenness in their boundaries (pull-apart basins).

4.8. The Location of Back Arc Rifting

The experiments shown that tensile stress occurring in the overriding lithosphere due to suction between the plates is insufficient to break apart the lithosphere if it does not contain a weak zone. Consequently, the presence of such weak zones must be supposed. Their location should determine the places where active back arc basins will form. Several possible causes for the lithospheric weakening can be considered, the first of which is volcanic activity. The lithosphere beneath a volcanic arc should be weakened due to the presence of volcanic conduits, magma chambers, and ascent of isotherms. The second mechanism for weakening is induced convection in the corner between the subducted and overriding plates. Considerable thermo-mechanical erosion of the base of the overriding plate can be produced by this process [*Hsui and Toksoz*, 1979], although the convection itself is insufficient to cause failure in the plate [*Artushkov*, 1979]. The thinning of the base of the overriding plate in experiment 4.5 (Figure 4.6) can thus simulate the result of either induced convection or volcanism. It is also important to note that the model lithosphere possesses approximately the same strength for both compression and tension. In contrast, real lithospheric material (rocks) is considerably weaker under tension than under compression. A model lithosphere possessing the same properties would require much less thinning (weakening) of the overriding plate to cause its failure.

According to the mechanisms for weakening of the lithosphere, its failure and initiation of an active back arc basin can develop either directly at the volcanic arc, (Figure 4.12*a*) or behind it (Figure 4.12*c*). The lithosphere breaks where it is weakest. The weakening effect of secondary convection depends on the subduction rate, the Benioff zone dip angle, the duration of subduction and other factors. It is known that volcanic activity also depends on some factors, such as the duration of subduction and the presence of topographic rises on the subducting plate. Subduction of these features can result in the complete cessation of volcanism [*Kelleher and McCann*, 1976].

Different locations for the formation of a back arc basin can be found in nature [*Taylor and Karner*, 1983]. The Mariana and Tonga-Kermadec arcs are classical examples of subduction zones where rifting occurred directly in the volcanic arc [*Karig*, 1972] (as is the case in Figure 4.12*a*). Active back arc basins are located now between the active and the remnant volcanic arcs. The northern Okinawa Trough [*Lee et al.*, 1980; *Letouzey and Kimura*, 1986], the Bransfield Strait [*Forsyth*, 1975] and the "Rocas Verdes" basin of southern Chile [*Dalziel*, 1981] are examples of lithospheric failure in the back arc area (Figure 4.12*c*). A third variant, the failure of the overriding plate in the fore arc area also seems to be possible. In fact, this case corresponds to a separation of the subducting and overriding plates (Figure 4.12*b*). Such a process probably has taken place at the beginning of opening the South Sandwich active back arc basin [*Shemenda and Grocholsky*, 1986] (see Sections 4.16 and 4.17).

Variation of the crustal thickness H_c of the overriding plate also influences the location of the plate failure under tension. Usually H_c is at a maximum right in the area of the volcanic arc, which enhances the probability of rifting in that area. However, the probability of the rifting in the back arc zone would increase if the overriding lithosphere is continental with a thick crust.

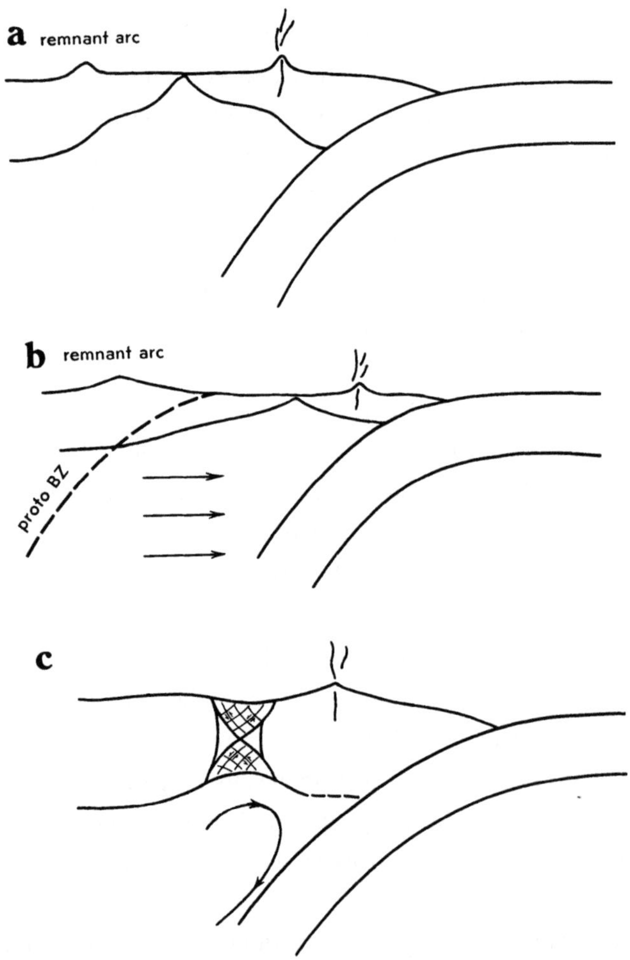

Figure 4.12. Back arc spreading initiated through (*a*) rifting of the volcanic arc, (*b*) failure of the overriding plate in the region of the inner trench slope (separation of the plates), and (*c*) failure of the overriding lithosphere behind the volcanic arc. BZ is Benioff zone.

4.9. Relationship of the Back Arc Regime to Relief and the Gravity Field in Subduction Zones

A subduction zone is shown schematically in Figure 4.2 for the two cases, $\rho_l = \rho_a$ (dotted line) and $\rho_l > \rho_a$ (solid line). The figure shows both a nonisostatic elevation of the overriding plate in the frontal arc region and a relatively flat dip of the Benioff zone corresponding to condition $\rho_l = \rho_a$. The back arc stress state in this case is characterized by across-arc compression. The subduction of the denser lithosphere causes subsidence in the fore arc area, a steeper Benioff zone dip, and back arc tension.

The rising of the frontal arc is due to superhydrostatic pressure σ_n from the subducting plate. On the other hand, greater density of the sinking plate causes subsidence of the overriding wedge. The first factor (tectonic pressure) is quasistationary while the second factor continually changes as the subducted plate is lengthened. These two competing factors define the frontal arc level in each case. The rising of the frontal arc is purely nonisostatic as it is caused by excess, nonhydrostatic pressure. Therefore, the well-pronounced elevation of the overriding plate for $\rho_l = \rho_a$ (Figure 4.2) must be manifested in the anomalous free air gravity field Δg as a distinct maximum above the frontal arc. Its amplitude was shown in Section 3.7 to amount to a few hundreds of milligalls.

The subsidence of the overriding wedge below the level of its isostatic equilibrium for $\rho_l > \rho_a$ (Figure 4.2) is compensated for by the underlying accumulation of denser material subducted into the mantle. The positive gravity effect of this mass is characterized by both a comparatively low amplitude and a large wavelength of the order of 1000 km [*Watts and Talwani*, 1975]. At the same time, sagging in the relief caused by this mass due to suction between the plates produces a negative anomaly Δg whose wavelength is several times shorter. Therefore, the subsidence of the overriding wedge must also be recorded in the gravity field as a local minimum. It follows thus that the anomalous gravity field Δg in a subduction zone is a diagnostic feature of the back arc regime. If maximum Δg marks the frontal arc and reaches the high values of 200 to 400 mGal one can expect the back arc area to be dominated by compression (Figure 4.13*a*). A volcanic arc usually has the highest hypsometric position in a subduction zone. Nevertheless, Δg above it frequently has only a local maximum of no more than 150 mGal. (This argues for the volcanic arc rise to be largely isostatic in origin.) If Δg drops at the frontal arc and the local gravity maximum at

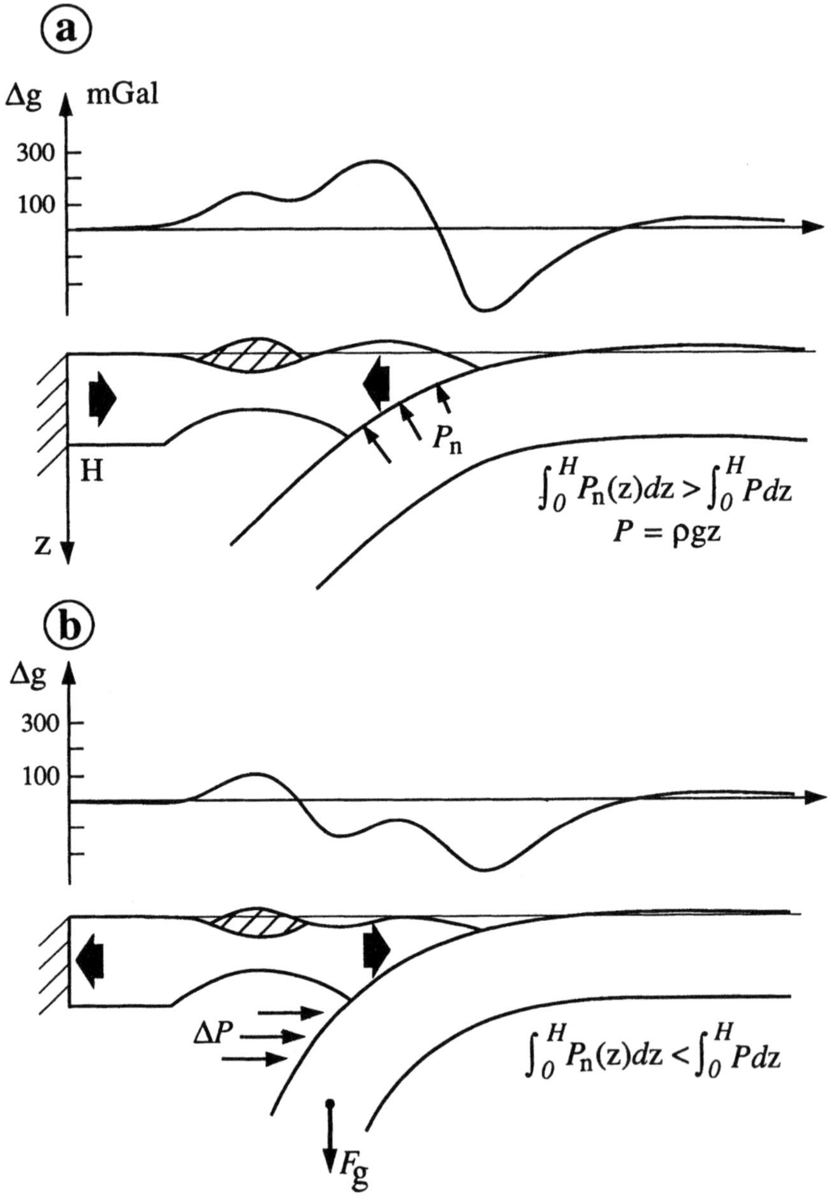

Figure 4.13. Relationship between anomalous free air gravity field and back arc stress regime. F_g is the weight of the subducted plate (the pull force)

the volcanic arc becomes an absolute maximum, then according to our results (Figure 4.13*b*) this would indicate back arc tension and rifting. To verify this conclusion it is necessary to compare the relevant data for Marianas- and Chilean-type subduction zones characterizing by back arc tension and compression, respectively [*Uyeda and Kanamori*, 1979]. Figure 4.14 demonstrates that data for the Kuril (Chilean type) and the Mariana arcs agree well with the prediction of the model (for comparison see Figure 4.2). These conclusions drawn for variable density of the subducting plate also remain valid for the other mechanisms determining the back arc regime. All the mechanisms control the stress state of the overriding plate through pressure in the interplate surface. The pressure in turn defines nonisostatic vertical movements in the fore arc area.

The tensile stress, σ_h, in the overriding plate reaches a maximum during the early stages of rifting and then decreases after breakage of the lithosphere and initial spreading. The larger the tension, the greater the drop in Δg that must take place in the fore arc area. The

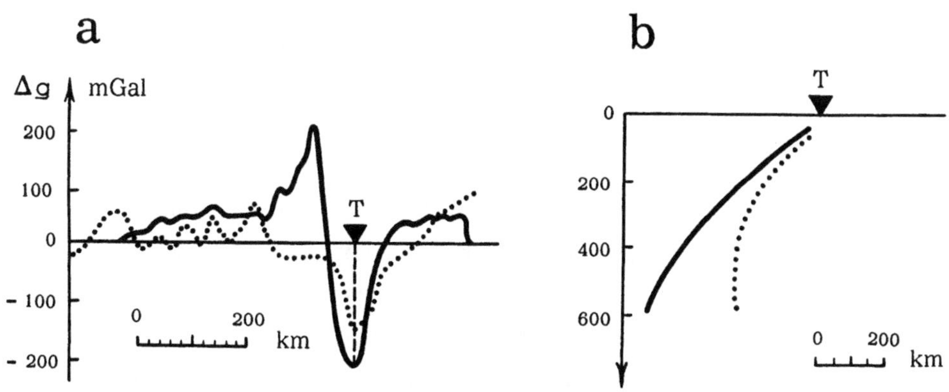

Figure 4.14. (*a*) Free air gravity anomaly Δg profiles and (*b*) sections of the Benioff zone surface crossing the Kuril arc at its central part [*Gnibidenko et al.*, 1980; *Kogan*, 1975] and the Mariana arc along 18° S [*Studies*, 1980]; the continuous line corresponds to the Kuril arc, the dotted line to the Mariana arc. T is the trench.

natural example where the overriding lithosphere has not yet completely broken under tension is the Okinawa Trough (rifting is still going on in some segments of this trough [*Letouzey and Kimura*, 1986]). As expected, the fore arc area in the Ryukyu subduction zone is characterized by a very low Δg, as low as -200 mGal (see deep linear gravity low along the Ryukyu zone in Figure 5.17; a trench in the relief which is also a trench in geodynamic sense ("subduction trench"), is shown in Figure 5.7 and is located to the south of the gravity minimum). Another example is the Aegean subduction zone where the overriding continental lithosphere is presently actively stretching (see Sections 4.18). The gravity anomaly Δg has a value of less than -100 mGal over the Hellenic Trough which is within the fore arc area (the "subduction" trench is located further to the south), Figure 4.21. The North Luzon Trough (fore arc basin) is also marked by a low Δg value amounting to -100 mGal to the south of the Taiwan collision zone [*Bowin et al.*, 1978], Figure 5.17. Therefore our model predicts tension of the overriding Philippine Sea Plate in this area. Unlike in the two previous examples, the extensional tectonic regime in this case is not obvious, even though there are some seismologic indications for tension in the region in question between ~19° and 22°N [*Lin and Tsai*, 1981]. Further to the north, Δg rises rapidly responding an increase in lithospheric compression due to the collision between the Luzon arc and the Asian continent (see Section 5.5). In the Mariana zone the stage of the rifting and breakage of the overriding plate has been already passed. That is why the gravity minimum in the fore arc area is not large, only minus a few tens of mGals (Figure 4.14). In all subduction zones mentioned above, the maximum Δg value is located at the volcanic arc and is typically less than 150 mGal which is in agreement with the model.

The model in Figure 4.13 thus provides an explanation for the phenomenon of double trench observed in the gravity field in some subduction zones. One "gravity trench" marks the trench in the relief ("subduction trench") which corresponds to the front of subduction while the other, the "fore arc trench" does not. The subduction trench exists (is preserved) because the sediments filling the trench are dragged down under the overriding plate but there is no mechanism removing the sediments from the "fore arc trench". This trench is filled with sediments and forms the fore arc basin. In some cases (in the Ryukyu zone, for example) the negative gravity anomaly in the basin is even greater than over the subduction trench.

4.10. Benioff Zone Dip Angle and Back Arc Regime

All mechanisms for back arc tension we have discussed (except aduction, experiment 4.7, Figure 4.10) force the sinking plate (the Benioff zone) to dip at a steeper angle. Such a tendency is indeed corroborated in some cases (see Figure 4.14 for example). At the same time this is not the case in other subduction zones. For example, the Izu-Bonin and Tonga-Kermadec arcs with active back arc basins have Benioff zone dip angles of only 60° - 65° on average. This, however, does not contradict our results. Figure 4.6*e* shows for instance, that the motion to the right (or seaward) of the broken off island arc continues even after the sinking plate has come to a rest against the bath bottom. The Benioff zone flattens during this stage of the experiment and the rate of the island arc motion gradually decreases. The rate reaches zero with respect to the bath (in the "mantle" reference frame) when the rate of the arc plate with respect to the underthrusting plate becomes equal to the rate of piston (Figure 4.2) motion towards the arc. The island arc then reverses the direction of its motion and approaches the opposite margin of the back arc basin. Similar processes may affect the Benioff zone dip angle in nature. It can occur, for example, when the sinking slab reaches more viscous mantle layers, i.e. the transition boundary at the 670 km depth.

It should be taken into account that the dip angle of the model plate sinking under its own weight (Figures 4.5 and 4.6), rapidly increases in the experiments due to the very low viscosity of the model asthenosphere. Actually, a dynamic equilibrium corresponding to a Benioff zone dip angle differing considerably from 90°, can be sustained between the rate of plate sinking and back arc spreading. This is demonstrated by the next experiment.

Experiment 4.8. Conditions are the same as in experiment 4.5 (Figure 4.6). The only difference is that the two plates are not initially pressed against each other and that the clearance between them is filled by "asthenospheric" material (Figure 4.15*a*). In other words, the effect of hydrostatic suction between the plates is "switched off" in this experiment. The subducting plate spontaneously begins to sink from the left end, with its hinge and the trench moving to the right. The rate of this motion is related to the rate of sinking so that the plate segment subducting into the mantle retains a low dip angle. If suction in the subduction zone prevented the trench from retreating, then the subducted plate would assume a steeper slope. Increased mantle viscosity would decrease this effect.

One can consider some other factors controlling the Benioff zone dip angle including the dynamic force associated with induced asthenospheric flow. Such flow diminishes the pressure above the sinking plate and consequently the Benioff dip angle [*Stevenson and Turner*, 1977; *Yokokura*, 1981].

The dip angle of the subducted plate certainly reflects the local patterns of interaction between the lithosphere and the surrounding mantle. However, it must be taken into account that sinking plates are elements of the global mantle convection and must be consistent with its structure. A model of mantle flow induced by present day plate motions was considered by *Hager and O'Connell* [1978]. The orientation of downwelling currents at convergent plate boundaries was found to be fairly close to the actual geometry of the Benioff zones. This would imply that the dip angle of the sinking slab is entirely determined by the mantle currents. Of course, such a conclusion and the model itself encounter certain objections [*Hsui,* 1988]. Nevertheless, it should be considered.

It is also worth noting that due to the rigidity (at least partial) of the subducting lithosphere (it can be compared to an inextensible film) the Benioff zone dip angle in each place should be related to position of this zone in the adjacent areas as shown in the scale experiments by *Yamaoka et al.* [1986]. To summarize, there appears to be no direct, simple relationship between the back arc regime and the subduction dip angle.

4.11. On the Cyclicity of Back Arc Tension and Compression

Periods of tension and compression in back arc basins seems to alternate with a time period of the order of 10 Ma [*Zonenshain and Savostin*, 1979; *Nakamura and Uyeda*, 1980; *Yamano and Uyeda*, 1985]. Such a periodicity may be associated with different kinds of thermo-mechanical instability in the mantle [*Jurdy and Stefanick*, 1983; *Yuen and Schubert*, 1977], as well as with the global reorganization of plate motion and collision in remote regions [*Uyeda*, 1984; *Yamano and Uyeda*, 1985]. Some of the possible causes for change of back arc regime also follow from the modeling. In particular, the dense subducting plate can periodically break, either under its own weight (as in experiment 4.4, Figure 4.5) or under the dynamic influence of the mantle. Before breakage, tension can exist in the back arc basin, but after that, compression is established and then the cycle repeats again. Also, resting of the sinking plate against the more viscous layers of the mesosphere can lead to attenuation of the back arc spreading and to compression of the overriding plate. Both mechanisms yield an estimate of several million years for the

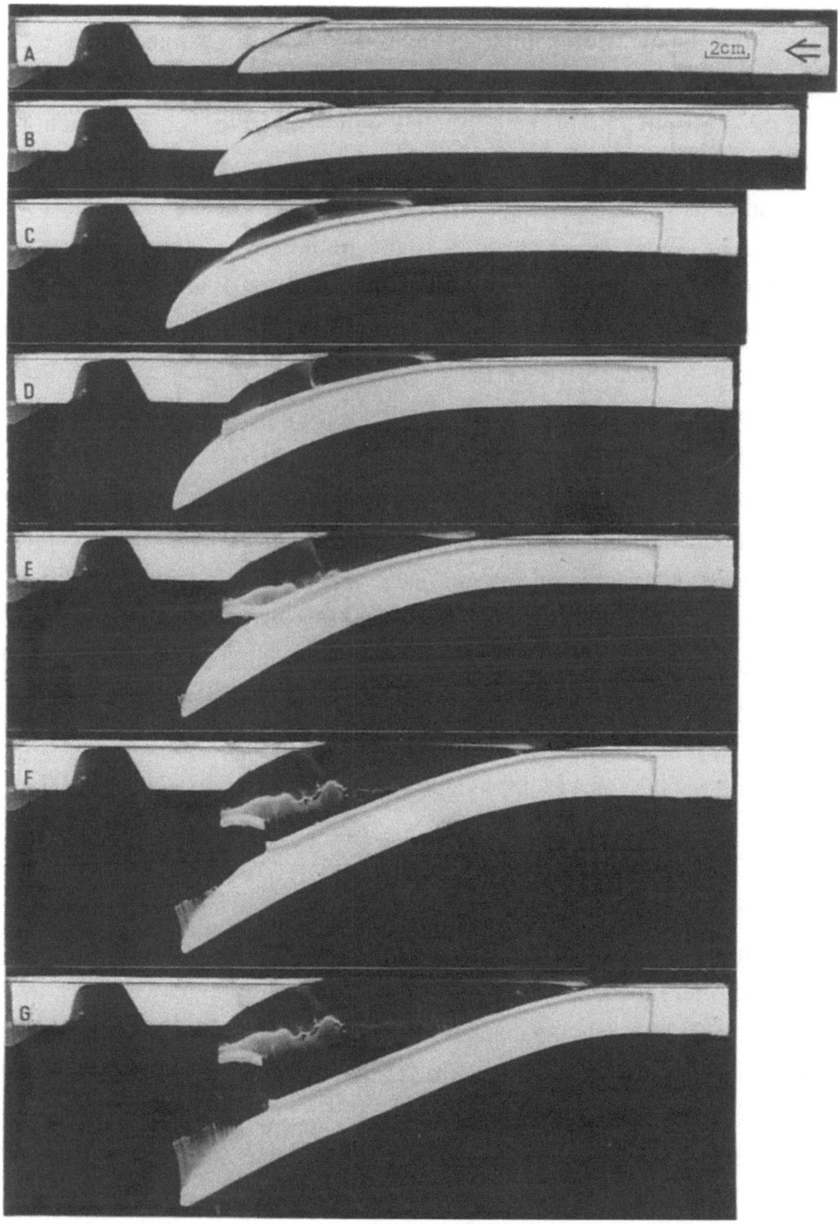

Figure 4.15. Experiment 4.8, spontaneous gravity sinking of the subducting plate under $\Delta\rho^{m} = 0.03 \times 10^{3}$ kg/m^{3} (white areas in Figures 4.15*e* to 4.15*g* are traces which the sinking slab leaves on the bath walls).

cyclicity of back arc regime.

The stress state of overriding plate changes also with a change in the age of the subducting lithosphere. The latter can be caused by jumps of the spreading axis during formation of the subducting lithosphere. Another important factor is the subduction rate V. Its increase in the model experiments retarded and completely stopped back arc opening. Under the conditions of experiment 4.5 (Figure 4.6) it was possible to select a rate of piston motion so as to make the broken off island arc immobile with respect to the bath. Reverse arc plate motion (with respect to the bath) and its collision with another fragment of the overriding plate occurred against the background of continuing subduction with further increasing V. The influence of V on back arc dynamics is controlled in the experiments by a change in friction and pressure between the plates in the subduction zone. They both increase with increasing V (the pressure grows due to the increase in force needed to bend the underthrusting plate at a higher subduction rate). The resultant direction and rate of the subduction hinge motion depend on the balance of horizontal components of friction and pressure forces on the one hand and the suction force which pulls the overthrusting plate oceanward, on the other hand. Under actual conditions the subduction rate affects the stress in the overriding lithosphere also through the viscous interaction of the sinking plate with the surrounding mantle. Viscous resistance to the subduction increases with growing V which contributes to increase in lithospheric compression (one can consider the resistance as a factor reducing the effect of excess weight of the sinking plate).

Considering the relationship between V and back arc regime it is useful for future reference to introduce the notion of "forced" subduction. Let us consider the following situation. The horizontal part of the subducting plate is fixed somewhere far from the subduction zone. Both subduction and back arc spreading develop only due to gravitational sinking of the oceanic plate into the mantle. The rates of these processes are hence the same in magnitude, but opposite in sign. If there is any additional external force (caused by mantle convection, for example) pushing the subducting plate toward the subduction zone, then the rate of back arc spreading would decrease. At some rate, this forced subduction would change the direction of motion of the arc plate resulting in compression of the opened back arc basin.

4.12. Vertical Movements in Subduction Zones

According to the modeling results a change in subduction mode and consequently in the back arc regime should be clearly reflected in vertical nonisostatic displacements of the arc plate. The amplitude of the movements in the model is of the order of 1 mm which yields for nature a value near 2.6 km. Figure 4.2 shows that a high level of the frontal parts of the overriding plate including the volcanic arc (the place where it is normally located) and its marginal sea slope corresponds to a compressional environment. The frontal arc undergoes maximum elevation. The elevation grows with increasing compression. Both decreasing compression and a change from compression to tension are accompanied by subsidence of the overthrusting plate including the volcanic arc area and, most of all, the frontal arc (see Figures 4.2 and 4.13). Similar patterns characterize actual subduction zones [*Fedorchenko and Piskunov*, 1977; *von Huene et al.*, 1980; *Uyeda*, 1982; *Sugi et al.*, 1983; *Letouzey and Kimura*, 1986]. A more complex sign-alternating vertical movements are observed in the model at the initial stages of subduction of a dense plate (Figures 4.5*a* to 4.5*e*). A stable subsidence of the frontal parts of the overriding plate develops after the initial stages have passed and quasi steady state subduction set down (Figure 4.5*f*). The picture of vertical movements also complicates when the compression of the lithosphere is very high and compressive stress approaches the yield limit σ_s in the arc area. Depending on the structure and evolutionary history of the arc, the overriding plate can undergo different types of flexural buckling which contribute to the pattern of vertical movements (see Section 5.3).

4.13. On Causes of the Arcuate Configuration of Subduction Zones

The curvature of the island arcs was initially believed to be a result of the Earth's sphericity. On the basis of zero strain buckling theory of spherical shells *Frank* [1968] and *Van der Held* [1969] proposed the models which relate the arc curvature to both the Earth's sphericity and subduction dip angle. This model however is not supported by the observations [*Tovish and Schubert,* 1978]. *Vogt et al.* [1976] have noticed that sharp change in the arc curvature is frequently associated with subduction of different topographic highs where the arc is as if pressed into the body of the overriding plate. It is exactly in this way that this observation was interpreted [*Vogt et al.*, 1976; *Hsui*, 1988].

The more buoyant parts of the subducting plate carrying the topographic rises were considered to be indented into the overthrusting plate causing the convergent boundary to bend. Physical modeling of this process conducted by *Grocholsky and Shemenda* [1985] has shown that the indentation (bending of the trench) does occur where a seamount or a ridge is subducted. However this process involves only the lower part of the inner trench slope. The width of the bent zone is comparable with the width of the subducting feature which normally is few to several tens of kilometers. In actual situation the "indentation zone" is much wider and involves the whole arc system including trench, inner trench slope, volcanic arc proper and back arc basin (Figure 4.16).

The modeling presented in this work suggests an alternative explanation [*Shemenda*, 1985]. The buoyancy of the lithosphere is indeed higher at the places where topographic rises are subducted. Its average density is lowered due to the lighter material of rises and their roots. Also, subduction of an asperity increases the friction between the plates in the subduction zone. Both these factors provide more intensive compression of the overriding plate or replacement of tension (if this existed before) by compression. If spreading was developing in the back arc zone then subduction of the ridge should result in slowing down or complete ceasing back arc opening. On the other segments of the subduction zone, back arc spreading may continue owing to non-rigidity of narrow arc plate (arc plate is broken by deep transcurrent fractures into blocks [*Carr et al.*, 1974; *Tarakanov and Kim*, 1980; *Lobkovsky*, 1988]). This would lead to the observed curvature of the whole arc system (Figure 4.16).

The subducting topographic rises thus do not press an island arc into the body of the overthrusting plate. Quite the opposite, these features retard the retreat of the subduction hinge to the ocean (trench roll back [*Dewey*, 1980]). It should be noted that this mechanism can not be regarded as universal. The regularities of lithospheric failure preceding the formation of subduction zone also play an important role. The configuration of forming subduction zone is strongly influenced by lithospheric inhomogeneities (see Section 3.4.1).

4.14. Relation of Seismicity in the Subduction Zone with Back Arc Dynamics

It is well known that Marianas-type subduction zones have lower seismicity than those of the Chilean-type [*Kanamori*, 1971; *Uyeda and Kanamori*, 1979]. *Ruff and Kanamori* [1980] concluded that the energy of thrust earthquakes in subduction zones roughly

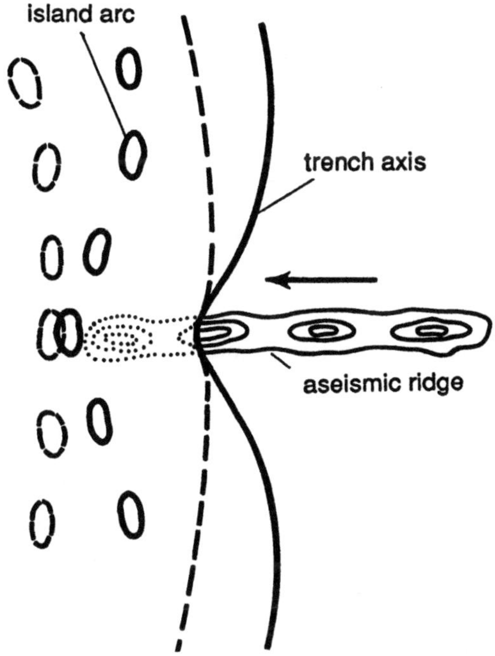

Figure 4.16. Bending of the island arc due to the cessation of back arc spreading at the places of subduction of an aseismic ridge. Dotted line designates the part of the ridge already subducted under the overthrusting wedge; dashed line indicates the position of the trench and the volcanic islands before subduction of the ridge.

depends on the age of subducting plate. *Peterson and Seno* [1984] have shown that to a first approximation the rate of seismic moment release per unit convergent boundary length is also a function of the plate age. These patterns may be caused by a different degree of coupling between the plates in subduction zone which supposedly decreases with increase in age of the subducting plate [*Uyeda and Kanamori*, 1979]. Using our results the coupling can be treated in terms of interplate pressure and friction. Decrease in contact pressure P_n (Figure 4.9*a*) results in change of the subduction mode from Chilean to Marianas-type. On the one hand, this leads to a direct decrease of friction τ_n between the plates, and on the other hand produces a "sucking" pressure gradient. The least viscous mantle derivatives (magma) driven by this gradient can penetrate (percolate) in some places into the interplate contact zone. This would cause a further decrease in the friction (coupling) between the plates. Seamounts of volcanic origin on the inner slope of the Mariana trench [*Hussong and Uyeda*, 1982] may have formed by such upward magma percolation along the interplate zone [*Shemenda*, 1985]. This conclusion remains valid also for other considered above mechanisms which control the lithospheric stress regime in the subduction zone through the contact pressure P_n.

4.15. On the Mechanism for Back Arc Tension in Some Particular Subduction Zones

We concluded above that there are several mechanisms which can change the interplate pressure P_n and thereby the back arc stress regime, the gravity field, and vertical movements in the subduction zone. In terms of this characteristics of subduction, the particular mechanism which operates in particular region and produce back arc tension (or compression) is indistinguishable. The situation can sometimes be clarified using the kinematic data on plate (microplate) motions in the "absolute" (hot-spot) reference frame (provided this frame is indeed absolute). *Carlson and Mortera-Gutierrez* [1990] have shown that in the Izu-Bonin and Mariana zones all interacting plates, including the Pacific (subducting), the Philippine Sea, and the arc (Mariana) Plates move westward with different velocities. The Philippine Sea Plate moves faster than the arc plate but slower than the Pacific Plate. Back arc spreading thus develops due to the arc plate (the whole subduction zone) lags behind the Philippine Sea Plate. Hence the main mechanism in this area seems to be an anchoring of the subducted plate. This conclusion is at some variance with previous ideas [*Molnar and Atwater*, 1978; *Uyeda and Kanamori*, 1979; *Uyeda*, 1984; *Shemenda*, 1985]. The opening of the Mariana Trough was thought to be caused by the subduction zone retreating to the ocean. The latter in turn was considered to be a result of gravitational sinking of the Pacific Plate which is corroborated by the facts that this plate is the oldest near the Mariana trench and that the Benioff zone is practically vertical. According to the mentioned kinematic data the subduction zone not only does not retreat toward the ocean but moves in the opposite direction at a rate of a few centimeters per year. If these data are true, then one can suppose that the Pacific Plate is actively pushed under the Izu-Bonin and Mariana arcs by external force, which means that subduction is forced (see Section 4.11). This force suppresses the retreat of the subduction hinge to the ocean caused by the suction force and moreover provides the advance of the hinge to the west.

The rate of trench (subduction hinge) advance should be higher when the component of its retreat is lower. This situation takes place at the locations of the subduction of topographic highs resulting in the same pattern of the arc curvature which discussed in Section 4.13 (Figure 4.16).

The situation in the Tonga-Kermadec arc is less well constrained. Opening of the Lau basin seems to be caused by retreat of the subduction zone. The overriding plate may also be moving in the same direction but with lower velocity [*Carlson and Melia*, 1984]. It is not quite clear what is the cause for trench retreat in this case. It can be either the asthenospheric currents or gravitational sinking of the subducting plate. Probably the most clear situation (in terms of constraining the particular mechanism for back arc extension) takes place in the South Sandwich subduction zone which is considered in detail below.

4.16. South-Sandwich Subduction Zone

To understand the processes developing in this zone one must consider the situation in the whole area between the South America and the Antarctic Peninsula. Figure 4.17 show that four main plates (South American, Antarctic, Scotia and Sandwich) interact in this area. The rates of relative motion of the South American, Antarctic and Scotia Plates are very small. In particular the net rate of left lateral shear along the southern and northern boundaries of the Scotia Plate is about 1 to 2 cm/yr [*Galushkin and Ushakov*, 1978; *Pelayo and Wiens*, 1989]. The fast relative motion of the Sandwich Plate is prominent upon this background. This plate is bounded to the east by the subduction zone and to the west by an axis of back arc spreading with a total spreading rate of about 8 cm/yr [*Barker and Hill*, 1981; *Forsyth*, 1975]. The small Sandwich Plate (microplate) thus is bounded by the practically immobile South American, Antarctic and Scotia Plates. In spite that this plate moves rapidly toward the Atlantic ocean forcing down the South American Plate in its path and forming an active back arc basin behind the arc. It could be expected that such a fast motion of the Sandwich Plate is caused by the subduction of a very dense portion of the South American Plate. However this plate is relatively young and has ages ranging from 25 Ma in the southern half of the South-Sandwich zone to 75 Ma in its northern part [*Forsyth*, 1975]. Hence, the subducting lithosphere as a whole can be even lighter than the underlying asthenosphere. The model of the anchoring subducting plate also does not "work" in this case. The Scotia Plate does move to the west in a hot spot reference frame. However, the rate of this motion is a mere 1 to 2 cm/yr [*Chase*, 1978] which is obviously

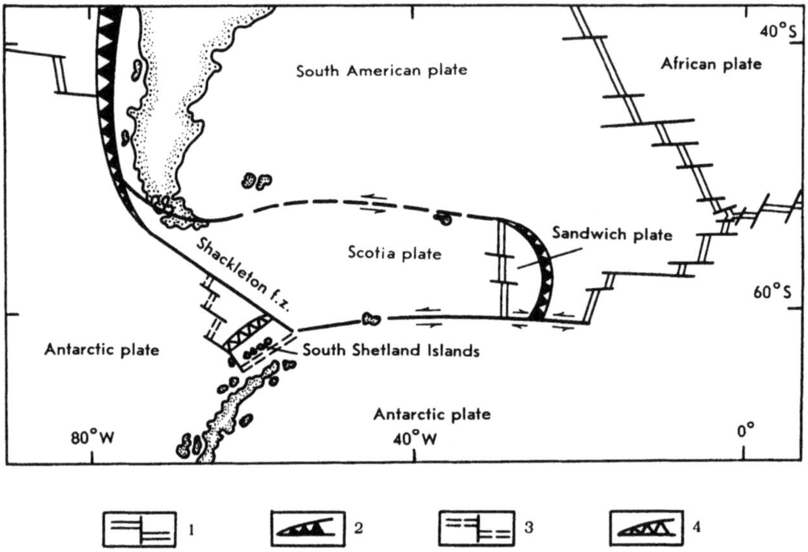

Figure 4.17. Geodynamic scheme of the Scotia Sea region [after *Barker and Hill,* 1981; *Forsyth,* 1975] 1, rift zones; 2, subduction zones; 3, inactive or nascent rifts; 4, ceased subduction zone.

insufficient to account for the fast back arc spreading. There is also no reasons to suppose that aduction (pulling-out) of the subducted plate occurs in the South Sandwich arc. Therefore it remains only to hypothesize that there is eastward asthenospheric flow under the Scotia Sea with a rate of the order of 10 cm/yr. Such a flow acts on the subducted segment of the South American Plate causing its eastward retreating [*Uyeda and Kanamori*, 1979; *Alvarez*, 1982; *Carlson and Melia*, 1984]. This results in an eastward motion of the arc plate sucked to the sinking lithosphere and in back arc spreading [*Shemenda and Grocholsky*, 1986]. The lithosphere subducted under the South Sandwich arc reaches depths of about 180 km and has a nearly vertical dip [*Frankel and McCann*, 1979] which is probably a consequence of the asthenospheric flow. Lithosphere of nearly the same age subducting under South America has a much flatter dip. More evidence for the asthenospheric flow under the Scotia Sea comes from gravity data (Figure 4.18). Free air gravity anomaly Δg can be treated as a result of the asthenospheric flow or more exactly of the excess (nonhydrostatic) pressure ΔP in the asthenosphere caused by the flow. To make a quantitative assessment of this hypothesis let us consider a kind of asthenospheric channel under the Scotia Sea. The lower boundary of the channel is a very viscous mesosphere, the upper boundary is the base of a thin oceanic lithosphere of the Scotia Sea, and the lateral walls are the North and South Scotia Ridges formed by thick continental blocks. The channel has the following dimensions: length $l_a = 3 \times 10^3$ km, width $d_a = 600$ km, height $h_a = 250$ km. The shape of the channel is something between a tube of diameter h_a and the plan-parallel layer of thickness h_a bounded by two infinite horizontal planes. To estimate the nonhydrostatic pressure in the asthenosphere for both cases we use a known solution of hydrodynamics [*Sedov*, 1976];

for the tube,

$$\Delta P = 12 l_a \eta V_a/(h_a)^2, \tag{4.11}$$

for the plane-parallel layer,

$$\Delta P = 32 l_a \eta V_a/(h_a)^2, \tag{4.12}$$

where V_a is the rate of asthenospheric flow; η is the asthenospheric viscosity; ΔP is the pressure difference at the inlet and outlet of the channel. For the natural channel we accept an average value from (4.11) and (4.12)

$$\Delta P = 22 l_a \eta V_a/(h_a)^2. \tag{4.13}$$

This nonhydrostatic (dynamic) pressure should cause a nonisostatic elevation of the overriding lithosphere (at the inlet of the channel with respect to the outlet) of the order of

$$\Delta h_g = \Delta P/\rho_a g, \tag{4.14}$$

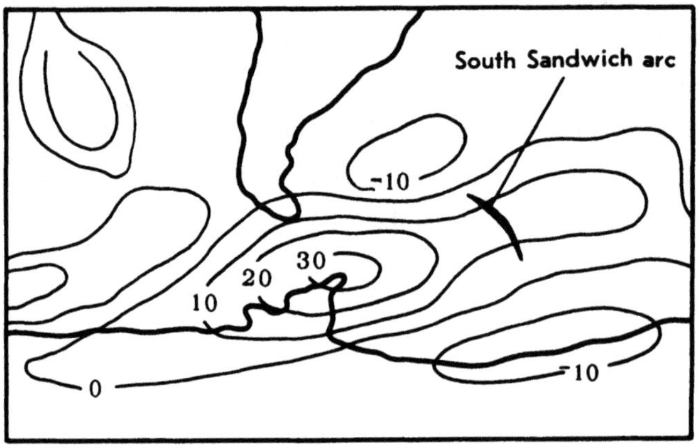

Figure 4.18. Anomalous free air gravity field Δg in the Scotia Sea area [*Ushakov et al.*, 1979].

where ρ_a is the density of the asthenosphere. The free air gravity anomaly Δg due to this nonisostatic elevation can be assessed as follows

$$\Delta g = 2\pi\rho_a G\Delta h_g, \qquad (4.15)$$

where $G = 6.67 \times 10^{-7}$ m/(kg $\times$ s^2) is the gravitational constant. Putting in this equation the observed gravity anomaly $\Delta g \approx 30$ mGal (see Figure 4.18) and indicated above values for other parameters we get $\eta = 3 \times 10^{19}$ poises. This is a reasonable estimate for the asthenospheric viscosity [*Artushkov*, 1979] which can be considered as corroboration of the existence of asthenospheric flow with the rate of the order of 10 km/yr under the Scotia Sea. The important point is that Δg reaches its maximum near the beginning of the channel in the region of the Drake Passage and then decreases eastwards. This confirms the suggested direction of the mantle flow.

What is the origin of the asthenospheric flow? This is a subject of special investigation which, however, is beyond the scope of this work. We only indicate some possible sources for the flow. One of the possible scheme is the following [*Alvarez*, 1982; *Shemenda and Grocholsky*, 1986]. Oceanic and continental lithosphere move with respect to the "slow", high viscosity mesosphere. Its top is at a depth of about 250 to 300 km. Between the oceanic lithosphere and the mesosphere there is a low viscosity asthenosphere. Under the thick continental lithosphere the asthenosphere is very thin or altogether absent. Within a framework of such a scheme the motion of the continents must generate the asthenospheric flow around them. The South American Plate moves westward in the mantle reference frame at a rate of about 2 cm/yr [*Minster et al.*, 1974; *Galushkin and Ushakov*, 1978; *Chase*, 1978]. The asthenosphere must be forced out from the high pressure zone occurring in front of the continent on the west and flow round it to penetrate in the zone of low pressure behind the continent. The asthenosphere can flow around the adjacent continents and through "windows" in the continental lithosphere like the Scotia Sea with a thin oceanic lithosphere. Taking into account that the Drake Passage is narrow, one can expect that the eastward asthenospheric flow under the passage is considerably faster than the westward motion of the South America. Under the Scotia Plate the accelerated asthenospheric flow moves between the vertical walls formed by the thick continental blocks on the North and South Scotia Ridges.

Another approach to the problem of origin of the asthenospheric flow under the Scotia Sea can be formulated in terms of global mantle convection. There are indications that large-scale convective upwelling mantle flow exists under the East Pacific Rise [*Monin and Sorokhtin*, 1979]. In the asthenospheric layer the flow spreads horizontally in different directions, including toward the South America. As in the previous case, this flow goes around the continent rushing to the "window" under the Drake Passage.

It should be noted that, no matter whether the upwelling flow under the East Pacific Rise exists or not, the lithosphere formed there moves toward the South America and is subducted under it. This motion itself induces the asthenospheric flow toward the South America.

Several lines of evidence are thus consistent with the existence of an accelerated asthenospheric flow under the Scotia Sea. The flow probably is the main mechanism for the opening of back arc basin in the South Sandwich subduction zone. It seems also that this flow and associated excess asthenospheric pressure played an important role in the evolution (opening) of the entire Scotia Sea region which is considered below.

4.17. Possible Mechanism for Evolution of the Scotia Sea Region

The transition zone between the South America and Antarctic continents started to form during the breakage of Gondwana in the Late Cretaceous [*Khain*, 1971; *De Wit*, 1977]. The modern configuration of South America, the Antarctic Peninsula, and the Scotia Sea formed during the last stage of the transition zone evolution which began 27 to 30 Ma ago [*Barker and Hill*, 1981]. This stage was characterized by general separation and relative shear of the South America and the Antarctic Peninsula. The high (excess) asthenospheric pressure on the Pacific side of the transition zone probably also played an important part. The pressure could be due to the same reasons as considered above, for example due to asthenospheric currents generated by the convergence of the Pacific and South America Plates. During splitting of these continents numerous continental slivers formed in the transition zone. High asthenospheric pressure facilitated separation of the slivers and pushed them to the east. Driven by the pressure the slivers overlapped the adjacent areas of the South American and Antarctic Plates resulting in expansion of the proto Scotia Sea (Figure 4.19). A low permeability for the asthenosphere of the northern and southern boundaries of the asthenospheric channel forming under the Scotia Sea was additionally maintained by the lithosphere subducted under the laterally spreading continental blocks. While the Drake Passage, the entire Scotia Sea and especially the marginal blocks of the Discovery Bank, Soth Orkney and South Georgia islands have been expanding (Figure 4.19*c*), the asthenospheric pressure under this region has been decreased due to eastward drainage of the asthenosphere through the opening passage into the asthenosphere under the Atlantic ocean. The drainage caused a drop in the asthenospheric pressure which resulted in ceasing both the opening of the Scotia Sea in the meridional direction and the subduction under the North and South Scotia Ridges. This is a time of initiation of fast asthenospheric flow (drainage) between the lithospheric roots of the continental blocks on the North and South Scotia Ridges. The asthenospheric flow acting on the plate subducted under the eastern boundary of the Scotia Sea produces bending of this plate, so that if this plate were fixed at the trench, the Benioff zone would oversteepen and dip towards the Atlantic ocean. However, the overriding plate at the eastern boundary of the Scotia Sea formed simultaneously with the subduction beneath it (Figures 4.19*c* and 4.19*d*). Consequently, at that time (the time of initiating of both the flow and back arc spreading)

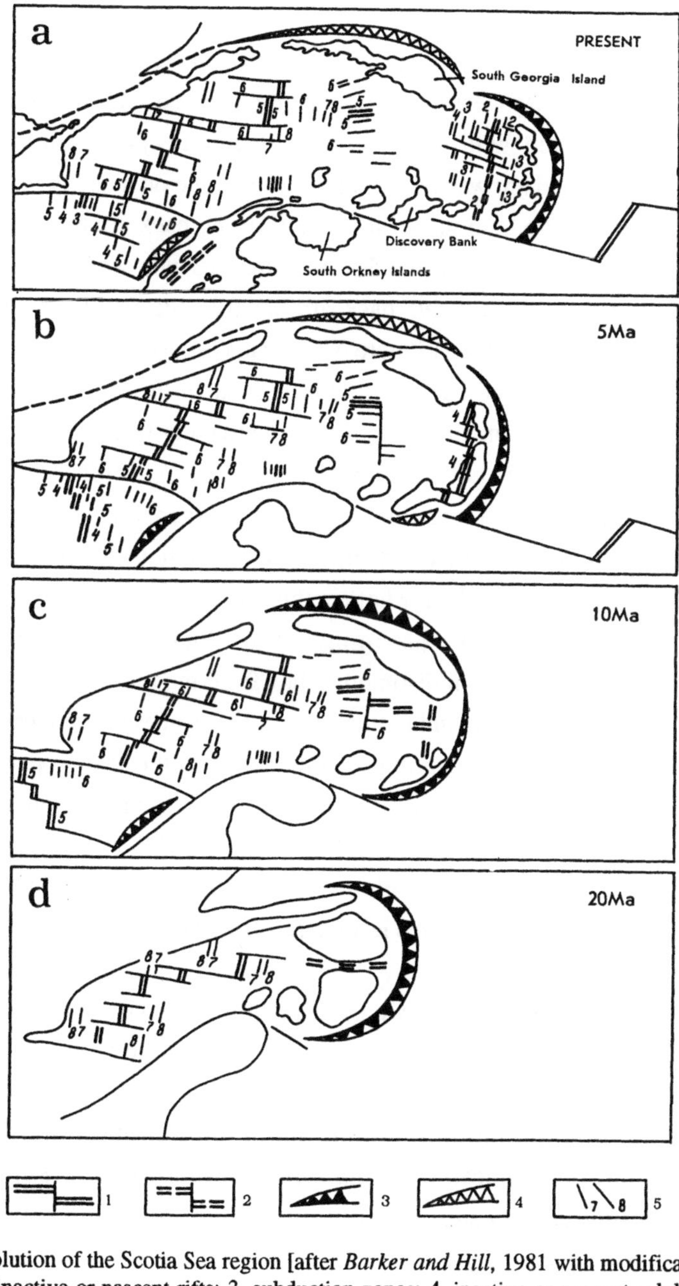

Figure 4.19. Evolution of the Scotia Sea region [after *Barker and Hill*, 1981 with modifications]
1, rift zones; 2, inactive or nascent rifts; 3, subduction zones; 4, inactive or nascent subduction zones; 5, magnetic anomalies.

this plate was very thin. Therefore, the subducting plate could be easily separated from the overriding lithosphere (perhaps tearing off only its foremost parts). The space between the spreading plates was filled with asthenospheric material crystallizing from the top downwards. The new island arc was initiated after some time on a newly formed crust and the old arc remained immobile to the west of the forming back arc basin (the Discovery Ridge connecting the South Georgia and South Orkney islands seems to be this remnant arc). Thus, we suppose that an active back arc basin in the South Sandwich arc was originated according to the scheme in Figure 4.12*b*. This is corroborated by paleomagnetic data showing that the South Sandwich arc or at least some of its islands have originated on oceanic crust 7 to 8 Ma old formed during back arc spreading which continues at present. The igneous rocks in the arc are not older than 4 Ma [*Barker and Hill*, 1981].

One of the predictions of the model is that the expansion of the Scotia Sea is accompanied by subduction under its northern and southern boundaries (largely along the northern boundary). Let us consider whether this prediction is in agreement with the available data. Dredging along the eastern part of the South Scotia Ridge has revealed "fresh" tholeitic basalts which are indistinguishable from the rocks composing the South Sandwich arc. The basalts correspond to relatively earlier stages of intraoceanic island arc volcanism which corroborates the northward subduction in the eastern part of the South Scotia Ridge. Dating of the dredged rocks has shown that subduction could develop approximately in the range 20 to 12 Ma [*Hill and Barker*, 1980] preceding the back arc spreading in the South Sandwich arc (Figures 4.19*c* and 4.19*d*). A present day motion along the South Scotia Ridge represents left lateral strike slip with some lateral extension [*Pelayo and Wiens*, 1989].

The North Scotia Ridge is marked by a positive free air gravity anomaly maximum of 150 mGal. It is bounded to the north by a trench with an average sedimentary fill of 4 km [*Rabinowitz*, 1977]. Detailed seismic sounding has revealed that the northern part of the ridge is the front of an accretionary prism which developed during the southward subduction of the South American Plate [*Ludwig and Rabinowitz*, 1982]. According to these data, a very slow ceasing subduction continues in this area at present, producing a collision of the North Scotia Ridge with the Falkland Plateau.

The Scotia Sea evolution thus represents a special case of back arc spreading. In fact, the entire sea is opened back arc basin formed due to high asthenospheric pressure beneath it. Subduction under the boundaries of the basin is secondary in this case (at least at the initial stages of the Scotia Sea opening), being the consequence of the sea expansion.

4.18. The Aegean Region

The important role of asthenospheric effects in the back arc dynamics can be also illustrated by the geodynamics of the Aegean region (Figure 4.20). This region includes the Aegean subduction zone, the Aegean Sea and surrounding continental blocks. The Aegean convergent boundary seems to possess all features typical of a "normal" subduction zone. There is a Benioff zone dipping northward at an angle of 35° reaching a depth of 200 km [*Papazachos*, 1973]. The projection of the 150 km isobath of the Benioff zone onto the surface approximately coincides with the front of the active volcanic arc. This subduction zone has double trench (Figure 4.21). One of them, the Hellenic trench, is located 200 km south of the volcanic arc, while the other is ever farther southward, behind the Hellenic arc and the Mediterranean Ridge (Figures 4.20 and 4.21). The composition of the frontal segment of the overriding plate reflects the complex evolution of this region. It was once the northern passive continental margin of a Triassic-Jurassic oceanic basin. After several episodes of compression and subduction since the Late Cretaceous to the Late Eocene this passive margin was converted into an active margin [*Biju-Duval and Dercourt*, 1980]. The Hellenic arc is an old elevated accretionary prism consisting of a series of thrusts packages of different ages corresponding to different subduction episodes. The Hellenic trench corresponds in its tectonic position to a fore arc basin in "classical" subduction zones. The trench in the geodynamic sense, the subduction trench (the place where the subducting plate is first underthrust beneath the overriding plate) is located further southward [*Angelier et al.*, 1982; *Prozorov and Shemenda*, 1987] (Figures 4.20 and 4.21). A new accretionary prism, the Mediterranean Ridge, is forming in front of the subduction trench [*Angelier et al.*, 1982]. This trench is poorly manifested in the relief and the gravity field and in some places is entirely absent. The poor expression of the trench is a consequence of the collision between the arc and the African passive margin commencing in this area. A progressively thicker (harder to bend) African Plate is subducted under the Aegean arc. The level of subducting plate surface becomes higher when approaching Africa due to thickening both light crustal and sedimentary layers. The frontal parts of the overriding plate are also pushed upward by the underthrusting lithosphere resulting in a shallowing the trench.

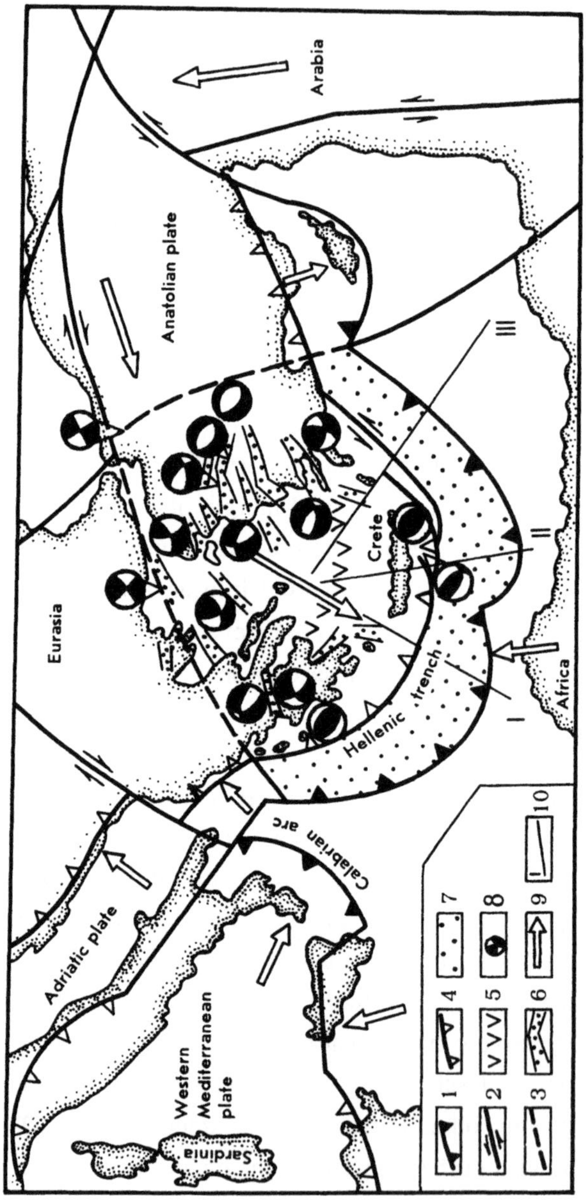

Figure 4.20. Geodynamic setting of the Aegean region [after *McKenzie*, 1972, 1978; *Morelli et al.*, 1975; *Ushakov et al.*, 1984; *Prozorov and Shemenda*, 1987]
1 to 3, plate boundaries (1, convergent; 2, transform; 3, poorly defined); 4, overthrusts; 5, volcanic (Aegean) arc; 6, Quaternary extensional structures (grabens); 7, Mediterranean Ridge; 8, fault plane solutions; 9, plate motion directions with respect to Eurasia (the arrows length is proportional to the rate); 10, location of the profiles shown in Figure 4.21.

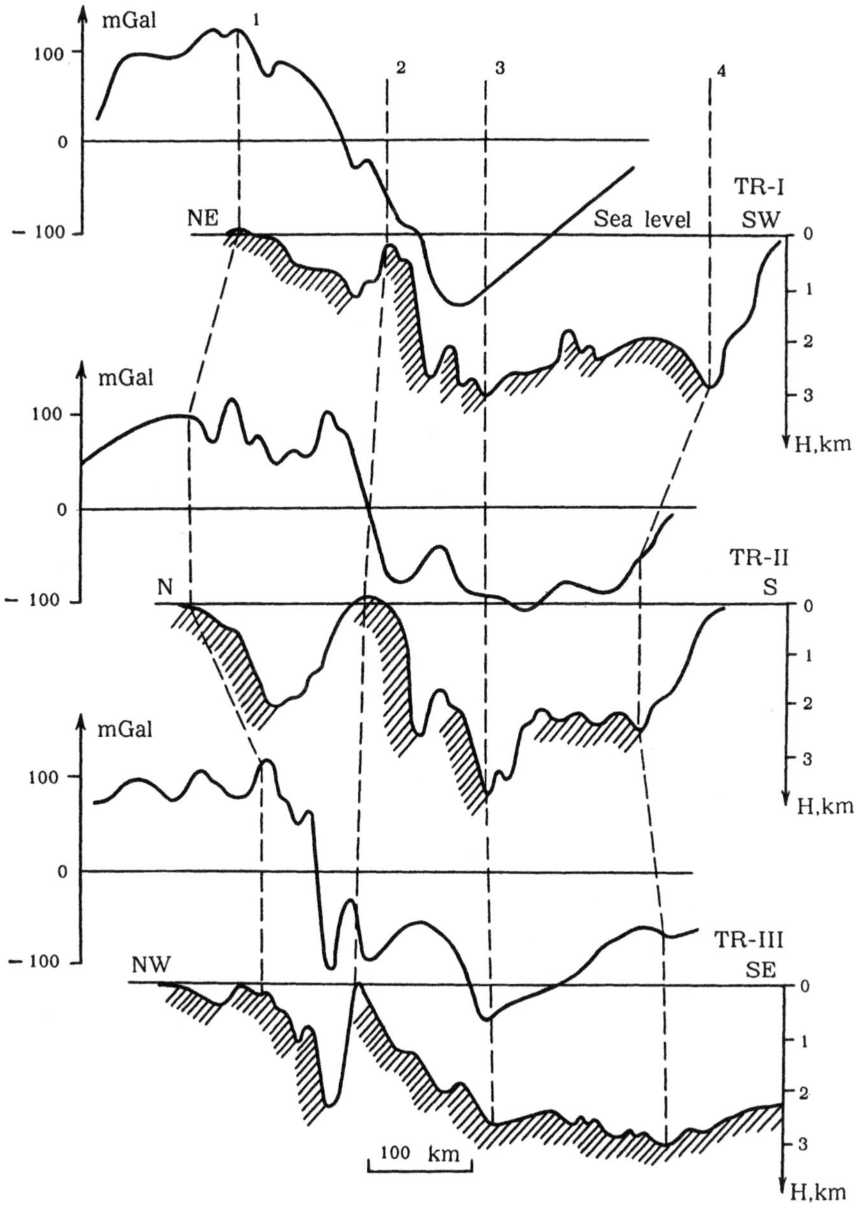

Figure 4.21. Bathymetric and free air gravity anomaly profiles through the Aegean zone along the traverses shown in Figure 4.20 [after *Angelier et al.*, 1982]
1, Aegean (volcanic) arc; 2, Hellenic (sedimentary) arc; 3, Hellenic trench; 4, "subduction" trench.

Collision causes considerable compression of the tectonic units along the entire convergent boundary. The compression is inferred from both fault plane solutions (Figure 4.20) and accretional tectonics in the frontal parts of the overriding plate. At the same time the lithosphere in the rear of the Hellenic arc under the Cretan and Aegean Seas as well as in western Turkey and eastern Greece undergoes extension in almost the same direction as plate convergence (Figure 4.20) [*McKenzie*, 1978; *Dewey and Sengor*, 1979; *Angelier et al.*, 1982].

Data on Raleigh surface waves, deep seismic studies and gravimetric data clearly indicate that the crust of the Aegean Sea is of the continental type, but is considerably thinner. Crustal thicknesses range from 22 km under the Cretan Sea to 28 km beneath the Cyclades archipelago to 32 km in Crete. In continental Greece and Turkey the crust reaches 48-50 km thickness [*Morelli et al.*, 1975; *Makris*, 1976; *McKenzie*, 1978; *Angelier et al.*, 1982]. Most investigators attribute this crustal thinning to horizontal nearly north-south extension of the lithosphere in the Aegean region amounting to a few hundred kilometers over the last 13 Ma [*Angelier et al.*, 1982]. What are the causes for such a strong extension which develops against a background of the beginning collision? Application of the models for back arc tension considered above does not yield any definite results. The most powerful mechanism for tension due to the asthenospheric (mantle) currents seems to be completely inapplicable to this region as it is surrounded by thick continental lithosphere. There is also no reason to assume a model for the anchored Benioff zone. Only the mechanism of back arc tension due to gravitational sinking of the lithosphere into the mantle remains plausible [*Meulenkamp et al.*, 1988]. The segment of the African Plate subducting in this zone seems to be old enough, being of 200 to 140 Ma [*Sendor and Yilmaz*, 1981; *Tapponnier*, 1977]. The depth of the Benioff (seismic) zone is about 180 km as mentioned before. On the other hand, according to seismic tomography data the subducted plate reaches a depth of at least 600 km [*Meulenkamp et al.*, 1988]. Even if the subducted plate is so long it is difficult to reconcile that its gravitational sinking alone is capable of causing considerable back arc extension in the Aegean region. This mechanism is comparatively weak (see Section 4.2) and furthermore, extension develops on the background of the beginning collision. The importance of "gravitational sinking" appears still more doubtful if one takes into account the small width of the Aegean subduction zone. These reasons prompts us to look for other causes of back arc tension in the Aegean region.

It seems that there is excess pressure ΔP in the mantle underlying the Aegean region. This is indicated by regional free air gravity anomalies which reach average values of 40-

50 mGal in the Aegean Sea and western Turkey [*Sazhin*, 1975]. The pressure can act on the subducted segment of the African Plate producing extension in the overlying lithosphere due to hydrostatic suction between the plates. The origin of the pressure may be associated with the kinematics of the lithospheric blocks directly in the Aegean area. The blocks are driven by external forces defining by general convergence of Africa and Europe. The asthenosphere in the areas where it exists or where its top is elevated should respond to the motion of the surrounding thick lithospheric blocks. If the asthenospheric lens undergoes horizontal contraction (due to the convergence of the thick surrounding blocks) it would result in growth of ΔP and flowing (squeezing) out of the asthenospheric material from zones of high pressure (zones which are shortened) into low pressure zones.

Seismological data indicate that there is a pronounced asthenospheric layer under the Aegean Sea. Its top is at 50 to 70 km depth [*Markis*, 1976; *Morelli et al.*, 1975]. The lithosphere of the Adriatic Plate is estimated to be 130 km thick in the Alpine area and 100 km thick in the Adriatic Sea [*Muller and Lowrie*, 1980]. The continental blocks on the western and eastern sides of the Aegean Sea are characterized by normal heat flow and seem to have even thicker lithosphere. We may therefore sketch the structure of the lithosphere in the Aegean region as shown in Figure 4.22.

There is a number of small plates in the area in question between the "vise jaws" of the converging European, African and Arabian Plates. The Arabian Plate is moving northward relative to Europe faster than the African Plate and pushes the small Anatolian Plate westward. The small Adriatic Plate is moving north-east, whereas the western Mediterranean and the Calabrian Plates are moving south-east. As a result, the thin lithosphere of the Aegean Sea which is trapped between the thick blocks of Turkey and Greece is undergoing east-west compression and shortening. Due to this shortening the underlying asthenosphere is being squeezed out resulting in increasing pressure ΔP. The asthenosphere tends to spread in all directions from the zone of high pressure, including toward the south where it is blocked by the Benioff zone which takes up the excess asthenospheric pressure (Figure 4.22*b*). It is this pressure along with gravitational sinking of the lithosphere into the mantle which probably produce the extension in the Aegean region.

It should be noted that there are other models for extension in the Aegean region [*McKenzie*, 1972; *Tapponnier*, 1977; *Dewey and Sengor*, 1979; *Mercier*, 1977; *Papazachos*, 1976]. They are based on the idea of horizontal south-southwestward extrusion of the Aegean Sea lithosphere produced by the mentioned above east-west

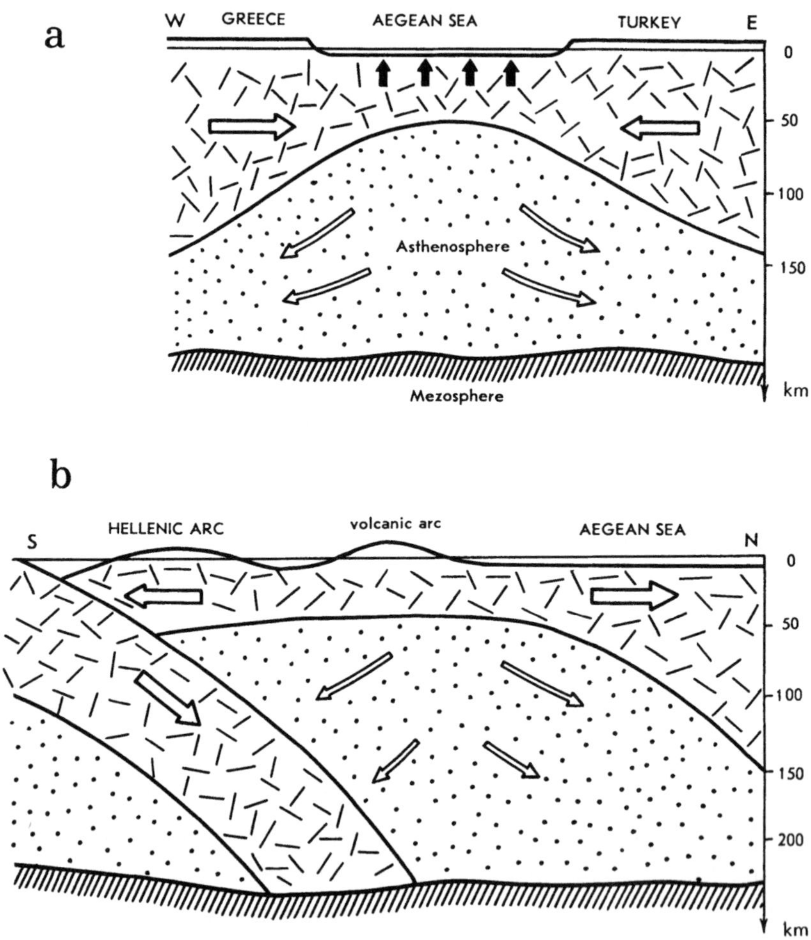

Figure 4.22. Mantle structure and dynamics in the Aegean region.
(*a*) East-west cross-section, (*b*) North-south cross-section.

compression of this area. The north-south lengthening of the Aegean Sea lithosphere results in its overriding onto the African Plate. In one case (the models by Tapponnier and Mercier), plastic extrusion of the Aegean Sea lithosphere develops along a system of slip lines. In another case (models by McKenzie and Papazachos) the meridional lengthening of the Aegean Sea is provided by compression of a mosaic of rigid plates in perpendicular direction. All these models explain certain regional features and the kinematics of the Aegean region, but grossly fail to explain the stretching and thinning of the Aegean Sea crust by a factor of 1.5 to 2. The area and hence the thickness of the crust are implied by these models to be constant to a first approximation (lengthening in one direction and thinning of the crust are produced by its shortening (and thickening) in the perpendicular direction).

Section 4.9 provides one more important criteria to test the applicability of different models for the Aegean region. This is pattern of gravity field which depends on the mechanism. If the overthrusting plate in the Aegean zone were ejected by external forces onto the subducting African Plate, the frontal (overriding) wedge of the Aegean Plate would have been elevated above its isostatic equilibrium level in the area of the frontal arc (see Figure 4.23*a* which is a modification of Figure 4.13). This would cause a high positive free air gravity anomaly Δg in the frontal arc. In other words, in this case the Aegean convergent boundary would be of Chilean-type. In fact, the maximum Δg is located in the volcanic arc and does not exceed 130 mGal (see Figure 4.21). This corresponds to a Marianas-type subduction zone (Figure 4.23*b*) and to the mechanism for back arc extension due to suction force as is suggested in our model.

The proposed mechanism can be applied to other pre- or syn-collisional situations where relative motion of thick continental blocks separated by contracting areas of a thin oceanic or continental lithosphere takes place. A differential pressure must inevitably occur in the asthenospheric lenses between thick lithospheric blocks. Such a pressure in turn should influence the regional motion of the lithospheric units. One can suppose the opening of the Tyrrhenian Sea in the Miocene-Pliocene time to have occurred due to such mechanism. This sea is at present bounded on the east by the Calabrian subduction zone and is pressed from the north and south between the Adriatic and African Plates (Figure 4.20). In the Middle Miocene, the Corsican-Sardinian continental block was situated near the Adriatic continental plate, which had a north-south orientation at that time [*Dewey and Sengor*, 1979]. When this block was subjected to pressure from the Eurasian and African Plates it

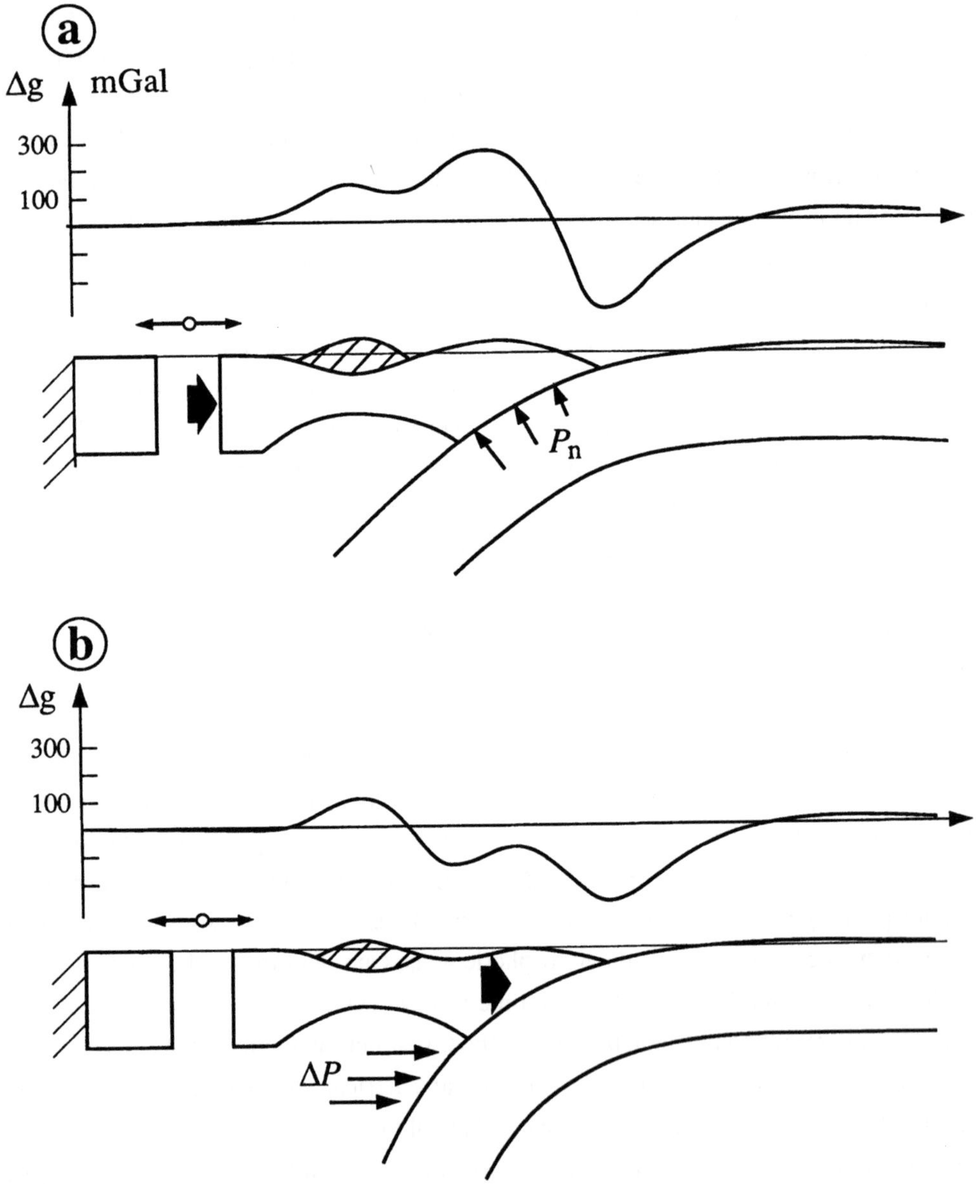

Figure 4.23. The gravity field pattern for two mechanisms of lithospheric extension in the Aegean region (see text).

was compressed in a north-south direction causing the Adriatic Plate to rotate counter-clockwise and the internal Tyrrhenian Sea to open in a nearly east-west direction. This east-west swelling of the Tyrrhenian Sea, combined with a north-south contraction, was probably caused by the hydrodynamic pressure effect in the underlying asthenosphere.

The final phase of the closing of the paleo ocean basin in the Carpathian-Pannonian region in the Miocene also seems to have involved a similar asthenospheric effect as is proposed for the Aegean region. The lithosphere of the basin was subducted under the Carpathian arc while the overriding plate was subjected to a north-south compression which resulted in formation of the Pannonian basin. This basin is also characterized by high Δg anomalies and a thinned lithosphere.

4.19. Conclusions

The main conclusions of this chapter are the following:

1. Tension of the overriding lithosphere arises in situations where some forces tend to separate the plates in the subduction zone. Such conditions can occur when (1) there is subduction of an old (dense) plate, (2) a horizontal asthenospheric (mantle) flow acts on the frontal surface of the subducted plate (Benioff zone), (3) there is a motion component of the overriding plate away from the subduction zone in the mantle reference frame, and (4) the relative rate of plate motion in the subduction zone is negative in sign, that is, when the subducted plate is being pulled back, towards the ocean basin.

2. Separation of the plates in the listed cases does not occur due to hydrostatic suction between the plates. The suction is caused by intensive hydrostatic pressing of the plates against each other in the subduction zone. To separate the plates, it is necessary to overcome the pressure, although this usually proves impossible due to the relatively low lithospheric strength (yield limit) σ_s. In other words, large tectonic stresses needed for plate separation cannot be created in the lithosphere as they cannot exceed σ_s. Therefore failure of the lithosphere would occur under existing tectonic tension in some weakened place before the subducting and overriding plates could be separated.

A more general conclusion is that the upper mantle, including a lithosphere of comparatively low strength and an effectively fluid asthenosphere, represents a special mechanical system where $\sigma_s \ll \rho_l g H$ is valid on average. Owing to this condition, separation of the lithospheric plates or blocks along existing boundaries requires considerable tensile stress comparable with the yield limit of the plates. At the same time, relative shear of the plates or blocks does not require a large force, this being determined only by friction.

3. Normally, tectonic tensile stress σ_h which can be generated in the overriding plate is several times lower than the effective yield limit of the lithosphere σ_s. For example, σ_h reaches only $(0.1\text{-}0.2)\sigma_s \approx 0.4 \times 10^8$ Pa for mechanism of gravitational sinking of the lithosphere (for the overriding lithosphere 60 km thick). Higher σ_h cannot be achieved for this mechanism, because after the subducted plate reaches a certain length, it is broken under its own excess weight. Such a comparatively low tension can be easily suppressed by the friction between the plates. Since back arc tension still exists in nature, the friction is probably low in actual subduction zones.

4. The location where the overriding plate fails (is rifted) under tension is determined first of all by weakened zones in the lithosphere. Failure occurs in the weakest zone. There are several mechanisms that weaken the plate. Depending on their effectiveness, failure in the lithosphere can develop either along an active volcanic arc or on either side of it. Variations of the crustal thickness H_c of the overriding plate also play an important role in determining the place of rifting. The greater the H_c value, the higher is the probability of lithosphere failure at this place.

5. The stress state of the overriding plate is governed by a change in the contact nonhydrostatic pressure σ_n in the interplate zone. If the σ_n value becomes less on average than the hydrostatic pressure, then back arc tension arises and vice versa. On the other hand, σ_n affects the vertical level of the overlapping plate surface and its departure from isostatic equilibrium. In other words, σ_n should determine the anomalous gravity field Δg in the island arc relating it to the back arc regime. A large positive free air gravity anomaly maximum Δg_m of 200 to 400 mGal marking the frontal arc is inferred to be evidence of compression in the overriding plate. A shift of Δg_m toward the volcanic arc where the gravity anomaly does not normally exceed 150 mGal is a sign of back arc tension. The larger the tension σ_h, the greater the decrease in Δg_m that must take place. Stress σ_h reaches a maximum during the earlier stages of back arc rifting. Then tensile stresses decrease after the breakage of the lithosphere and initial spreading.

6. Subsidence of the fore arc area due to a decrease in σ_n can reach a few kilometers resulting in the formation of quite wide and deep fore arc basins or deep sea terraces like in the Luzon, Ryukyu, and some other subduction zones. The basin filled with sediments will persist also after the change in stress regime. This mechanism for formation of fore arc basins (deep sea terraces) differs from that considered in the previous chapter (Section

3.4).

7. A consideration of the geodynamics of the Scotia Sea and Aegean regions shows that an important (perhaps even principal) role in back arc dynamics is played by the dynamics of the asthenosphere. Both the viscosity and flow rate of the asthenosphere have proved in nature to be sufficient to essentially affect the lithospheric dynamics not only on a global scale, but on the regional level too. The differential pressure occurring due to asthenospheric flow on opposite sides of thick continental blocks seems to be sufficient (under certain conditions) to force these blocks to override a thinner oceanic lithosphere and to form an active back arc basin behind. On the other hand, the continental blocks driven by some external forces can change (increase, in particular) the pressure in the asthenospheric lenses confined between these blocks. This can in turn, cause extension of the overlying lithosphere perpendicular to compression of the asthenospheric lenses. Phenomena of this kind seem to contribute to the north-south extension of the lithosphere in the Aegean and some other regions.

8. The model experiments have shown that a process opposite to subduction (i.e., aduction) is physically quite realistic.

CHAPTER 5

JUMPS OF SUBDUCTION ZONES, SUBDUCTION REVERSAL, AND CLOSING OF BACK ARC BASINS

5.1. Introduction

In Section 4.11 we discussed the mechanisms for increasing compression of the lithosphere in a subduction zone. These mechanisms are; subduction of young lithosphere, an increase in subduction rate, subduction of different asperities (seamounts, ridges) and plateaux. The most intensive compression occurs during subduction of the continental margin, that is during collision of the subduction zone with a continent. High compressive stresses can produce failure of the lithosphere in a new place resulting in the initiation of a new subduction zone at this place. Such a phenomenon is recognised in many fold belts [*Dewey and Bird*, 1970; *Kovalev*, 1985; *Koltypin*, 1991] and can be designated as a jump of subduction zone (similar to the jumps of spreading axis in fast [*Shemenda and Grocholsky*, 1991], slow [*Shemenda and Grocholsky*, 1994] ridges and the jumps of continental rifts [*Malkin and Shemenda*, 1991]). With jumps of subduction zones are related such processes as reorganization of the plate boundaries, evolution of the inner trench slopes, reversal subduction [*Dickinson and Seely*, 1979; *Mitchell*, 1984], as well as closing of back arc basins. This chapter is devoted to these problems.

5.2. Location of the Lithospheric Failure and Jump of Subduction Zone

In the previous chapters we already considered two situations which result in a jump of the subduction zone. One is related to a cessation of the subduction resulted from large scale continent-continent collision (see Section 3.4.1). The India-Eurasia collision caused an increase in compression of the Indo-Australian Plate. The compressive stresses probably reached the yield limit of the oceanic lithosphere to the south of Hindustan and produced an intensive intraplate deformation in the Central Basin of the Indian Ocean. If the deformation continues in the same way, a new convergent plate boundary might form in this area in the future (Section 3.4.1) resulting in a jump of the subduction (presently collision) zone over a long distance, near 3000 km.

Another example of the subduction zone jump is represented by experiment 4.2 (Figure 4.3, see also Figure 5.1). Failure in this case occurs within the subducting oceanic lithosphere in a flexural sag behind the outer rise due to strong resistance for the subduction. Along with very low density of the subducting plate (as in experiment 4.2) such a resistance can be caused by subduction of an extensive plateau with a thick crust (as shown in Figure 5.1), by high viscosity of the mantle, or by the presence of "obstacles" in the mantle as in the situation shown in Figure 5.14*e* (see Section 5.5.4 for details). Such a process can develop only under very high compressive force. If the plate contains a weak zone, then lithospheric failure can occur in this zone at a much lower level of compression. Considering back arc rifting in Section 4.8 we concluded that failure of the lithosphere in the subduction zone in the extensional environment usually occurs in the volcanic arc

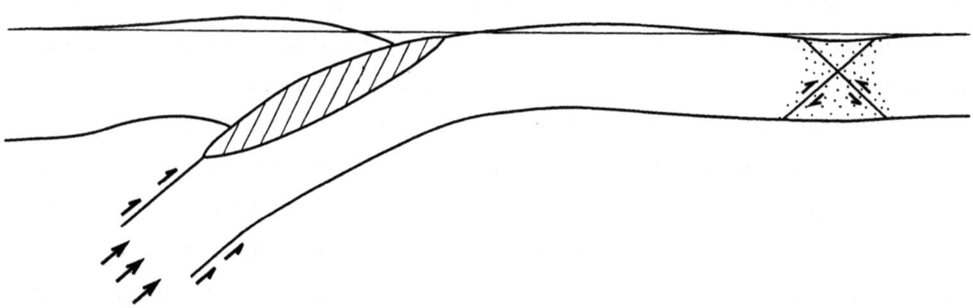

Figure 5.1. Jump of the subduction zone in the flexural sag behind the outer rise caused by subduction of the extended plateau or strong resistance from the mantle (see text).

where the lithosphere normally is the weakest. Similarly, one can suppose that in a compressional environment the volcanic arc is also the most favourable place for the lithospheric failure and for the subduction zone jump. Another likely location of the lithospheric failure is the back arc area where the lithospheric weakening can be produced by the induced convection (Section 4.8) or by the preceding back arc spreading (during the phase of back arc extension which had then been changed by strong compression).

5.3. Mechanism for Lithospheric Failure and Polarity of a New Subduction Zone

The lithospheric failure and the subduction zone in the indicated above locations develop through buckling instability, localization of deformation, and failure of the lithosphere along one of two oppositely dipping faults. For the first two situations presented in Figures 3.3 to 3.6, 3.15 and Figure 5.1, both possible directions of the failure seems to be equally likely. However, this is not the case when the failure of the lithosphere occurs not far from the subduction zone, in the arc or back arc areas. Below we present an experiment in which the overriding plate is thinned, imitating young and thin lithosphere of the back arc basin formed during the preceding stage of extension and spreading (Figure 5.2).

Experiment 5.1. The mechanical parameters of the model are the same as for experiment 3.1 ($\rho_l^m = \rho_a^m$), see Section 3.2. The lithosphere consists of two plates (Figure 5.2); a subducting and an overriding one, separated by an inclined cut (as in many previous experiments). An "aseismic ridge" parallel to the subduction zone is placed on the subducting plate. The ridge is made of compound N2 (see Section 2.3) which possesses nearly the same properties as compound N1 for modeling the oceanic lithosphere, but has a lower density 0.86×10^3 kg/m^3. The ridge was prepared by pouring the molten compound N2 into a special mould (without a bottom) placed on the plate surface. Such a "hot" technique ensures strong cohesion of the ridge to the subducting plate.

In Section 3.2 it was shown that to produce subduction at $\rho_l = \rho_a$ it is necessary to apply a compressive stress to the lithosphere equal to ~ $0.3\sigma_s$. This compression is insufficient however, to produce failure in thinned back arc lithosphere in experiment 5.1. The plate yield limit is reached at the stage of subduction of the ridge (Figure 5.2*d*) due to both increased friction between the plates and decreased average density of the subducting slab. Strong compression causes flexural buckling of the back arc lithosphere. Localization of deformation then develops in the sag nearest the island arc resulting in failure of the plate

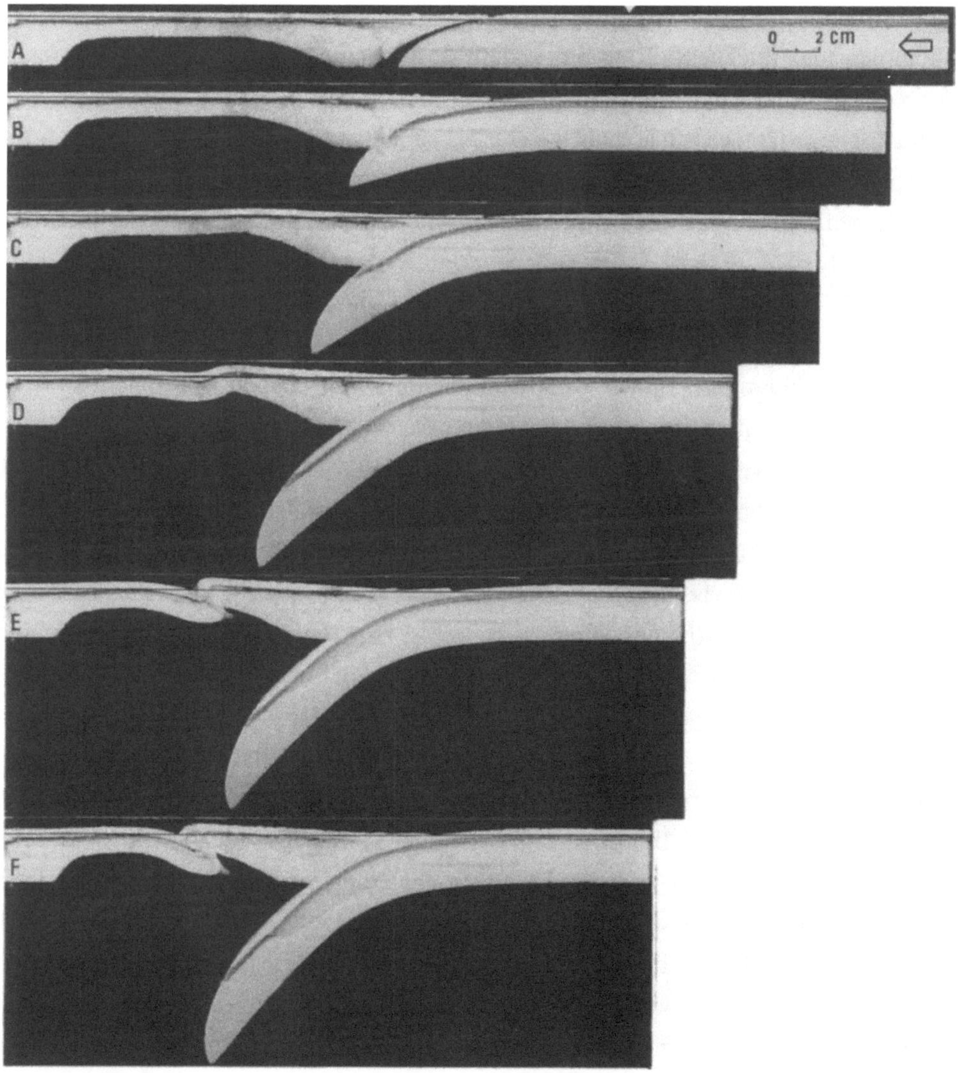

Figure 5.2. Experiment 5.1; subduction of a plate carrying an "aseismic ridge" parallel to the "trench" ($\rho_l^m = \rho_a^m$). The ridge is not seen in the photo, its position is shown by the white arrow in Figure 5.2*a*.

along an inclined fault dipping under the arc. Further compression produces subduction and closing the back arc basin. The buckling and localization of deformation in the overriding plate in this experiment is similar to those in other experiments with inhomogeneous thickness of the lithosphere (in particular, experiment 3.3, Figures 3.3 and 3.4) where buckling and failure occur in the vicinity of zone with plate thickness contrast. The difference is that in different trials under the conditions of experiment 5.1 the lithospheric failure always occurred along the plane dipping under the arc plate (under the trench), while in the previous experiments both opposite directions of the failure are equally likely.

Figure 5.3 presents a summary of the results of other experiments conducted under different conditions with a local thinning in the overriding plate corresponding to the volcanic arc. Figure 5.3*a* shows "usual" subduction of the oceanic lithosphere under the thinned overriding plate for $\rho_l^{\mathrm{m}} = \rho_{\mathrm{a}}^{\mathrm{m}}$. Compressive stress in the lithosphere in this case is considerably lower than the yield limit in the thinned zone. Under the pressure from the subducting plate, the overriding lithosphere is smoothly elevated in the frontal arc and less in the arc area. We also conducted a set of similar experiments with progressively higher compression. An increase in compression was achieved by increasing the thickness of the subducting lithosphere (higher compressive stresses are required to produce deformation of a thicker plate in a subduction zone). Increasing compression σ_{h} first causes simply an increase in the nonisostatic elevation of the overriding plate. The style of deformation however changes when σ_{h} approaches the yield limit of the overriding plate in the thinned zone (arc area), Figure 5.3*b*. This area undergoes some shortening and uplift due to the flexural buckling of the lithosphere. Two thrust fault zones then begin to develop. Complete failure of the plate occurs along the zone dipping under the trench (similar to experiment 5.1), with a new subduction initiating in this zone. This dip of the fault and a polarity of the resulting subduction are stable and were repeated from one experiment to another. The result however changes when the thinning in the overriding plate is made far enough from the subduction zone (Figure 5.2*c*). First, the deformation in the thinned zone develops similarly to Figure 5.2*b*, but then resultant failure of the plate occurs along different zones in different experiments of this type. Basing on the conducted experiments it seems that the probability for the complete failure of the plate along both zones is more or less equal. A similar result was obtained in the experiments with simple horizontal compression of the lithosphere containing a notch, see Figure 3.2. In such experiments,

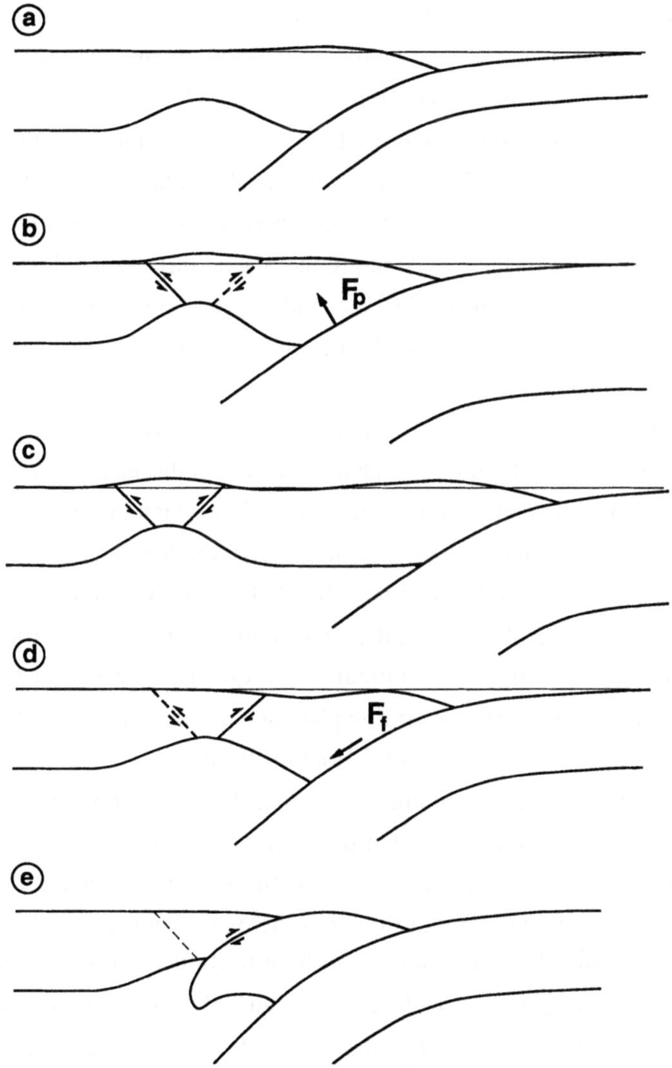

Figure 5.3. Deformation of the overriding plate thinned in the arc area under different conditions (see text) (*a*) Moderate compression of the lithosphere. (*b*) High compression caused by the subduction of a thick plate (compressive stress is close to the lithospheric yield limit in the thinned zone). (*c*) Conditions are the same as in the previous case except the thinned segment is displaced far from the subduction zone. (*d*) High compression caused by a strong friction between the plates. (*e*) Consumption (subduction) of the fore arc block.

the plate is subjected to a homogeneous compression which makes both directions of the plate failure equally likely. Which factors control the stable dip of the resultant fault under the trench in the case shown in Figure 5.3*b*? In this case a thinned zone is located near the end of the plate, the place of the load application where the stress field is not homogeneous. A nonisostatic pressure force F_p (Figure 5.3*b*) from the subducting plate has two components horizontal and vertical. The horizontal component produces compression of the whole overriding plate whereas the vertical component produces an uplift of the frontal wedge of this plate. Such an uplift is kinematicaly consistent with failure and displacement of the wedge along the plastic thrust zone dipping beneath the trench and is inconsistent with displacement along the other zone. When the thinned segment of the overriding plate is located far from the subduction zone (as in Figure 5.3*c*) plastic deformation and failure of this segment occur without "feeling" the way of the load application. Deformation develops in a homogeneous stress field that makes this case similar to experiment 3.2 in Figure 3.2. What is the critical distance between the weakened segment of the overriding plate (the volcanic arc) below which failure of the plate always occurs along zone dipping under the trench and above which both directions of the failure are possible? Obviously, this distance is controlled by the wavelength λ of the flexural buckling of the overriding plate which depends mainly on the lithospheric thickness and compressive stress (see Section 3.8). Normally λ is equal to a few hundred kilometers. An uplift of the frontal wedge of the overriding plate due to force F_p (Figure 5.3*b*) can be felt over a distance of approximately $\lambda/4$ to $\lambda/2$ from the frontal arc.

A resultant failure of the overriding plate along the thrust zone dipping in the same direction as the original subduction (to the west in Figure 5.3) can be also obtained even if the weakened segment of the plate is close to the subduction zone. This occurs, for example, when strong compression of the lithosphere arises not due to the high interplate nonhydrostatic pressure force F_p, but due to high friction τ_n between the plates. If the vertical component of the friction force F_t (Figure 5.3*d*) is larger than the vertical component of the pressure force F_p, then instead of rising the frontal block of the overriding plate experiences subsidence, being dragged down by the friction force (Figure 5.3*d*). This makes the failure of the plate along the zone dipping to the left kinematicaly more favourable (Figure 5.3*d*). The block (the arc plate) located between this fault and the subducting plate is then subducted under the arc (Figure 5.3*e*). Under the conditions of the described experiments (when density of the lithosphere is the same as density of the asthenosphere) the block can be subducted completely into the mantle resulting in a migration of the trench toward the arc. Such a process can be regarded as a discrete (catastrophic) erosion of the overriding plate resulting in a rapid advance (jump) of the volcanic front to the back arc area over a considerable distance (several tens of kilometers). Under other conditions, however, underthrusting of the fore arc area beneath the arc can only develop partially and then complete failure of the lithosphere occurs along a thrust zone dipping under the arc from the opposite side. Such a process probably is going on currently in the Taiwan area which is considered in detail below (Section 5.5).

Deformation of the lithosphere in the arc may also depend on the load of the arc (volcanics) itself. In the case shown in Figure 5.4*a* to 5.4*c* we have put on the surface of the thinned part of the plate an additional load, a lens of the material of the same or lower density (it is not important in this case) which represents the volcanic arc. Under the load of this lens the plate is bent down slightly (Figure 5.4*a*). Then homogeneous horizontal compression is applied to the ends of the plate which are far enough from the arc. An increase in compression causes further flexural subsidence of the arc (Figure 5.4*b*) in contrast to the case without the load of volcanics (Figures 5.4*d* to 5.4*f*) where compression produces flexural uplift of the thinned plate segment. Further compression of the plate with "volcanics" results in the localization of deformation in the arc area (Figure 5.4*c*) and in the formation of two intersected thrust fault zones. During this stage (which is similar to the process shown in Figure 3.4) the surface of the deforming arc area experiences uplift (Figure 5.4*c*). Complete failure then occurs along one of zones. Failure of the plate without "volcanics" develops in a different way shown in Figures 5.4*d* to 5.4*f*. If compression is produced by the subduction and if the arc is located near the subduction zone, then depending on the particular mechanism of an increase in compression, failure along one of these zones will be preferable (similarly to the previously considered cases in Figures 5.3). In the process shown in Figure 5.4*a* to 5.4*c* the volcanic arc (volcanics) itself may or may not be involved in deformation (shortening and thrusting) at the stage of deformation localization. It depends on the coupling between the arc and the underlying plate. If coupling is weak, then the arc experiences only vertical movements in response to the deformation below.

Which of two model approaches presented in Figure 5.4 (with additional load of the arc, Figures 5.4*a* to 5.4*c* or without it, Figures 5.4*d* to 5.4*f*) is more realistic? It depends on whether the arc is isostatically compensated or not (whether the nonhydrostatic stresses in the plate caused by its bending under the arc weight have relaxed or not). We conducted experiments (Figure 5.5) with an arc and a "crustal root" beneath it made of compound N2

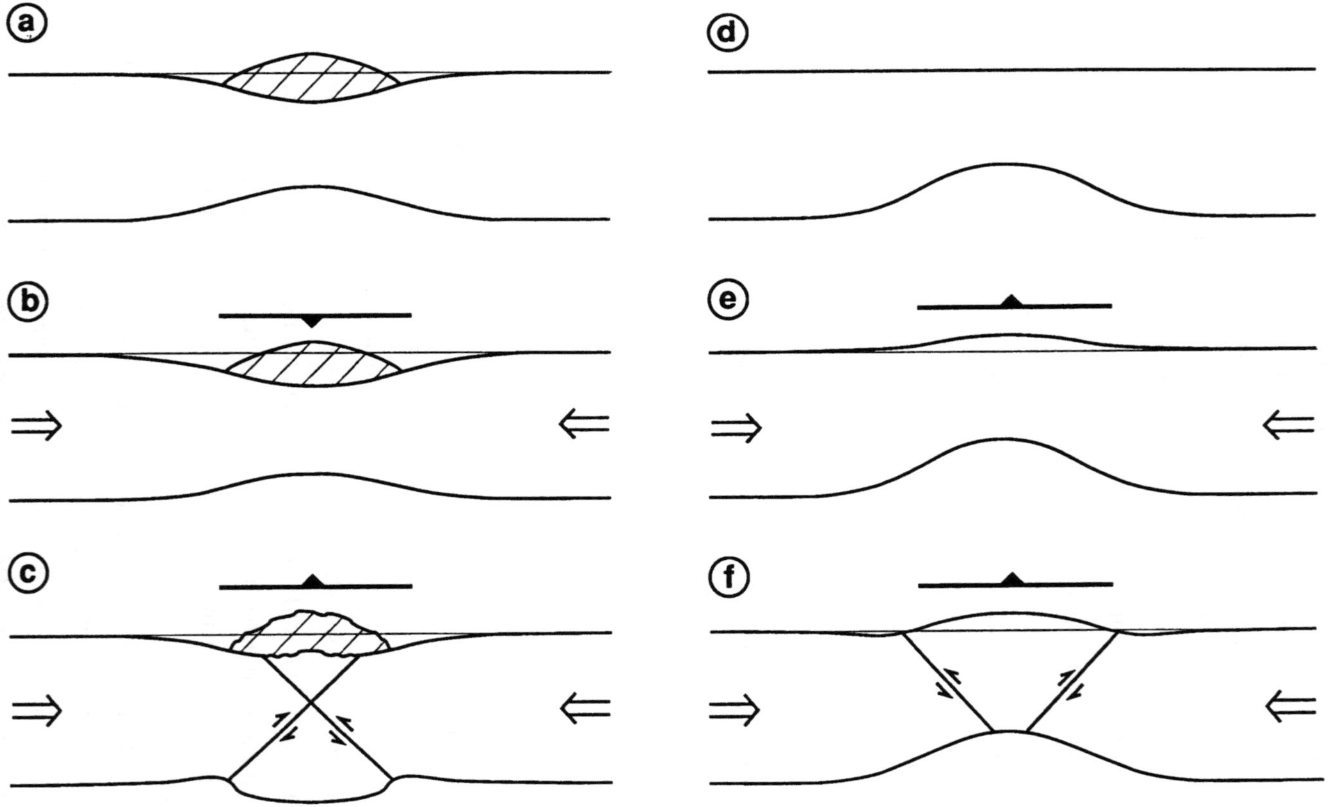

Figure 5.4. Deformation and vertical movements of the thinned lithosphere (Figures 5.4*a* to 5.4*c*) with an additional load due to the volcanic arc and (Figures 5.4*d* to 5.4*f*) without it.

which has a density less than the mantle, see Section 2.3. The depth of the root and height of the arc surface were made to fulfil the condition for complete isostatic compensation of the arc (there is no bending of the plate under the arc weight before compression, Figure 5.5). Compression of such a plate yields the same result as without the arc (i.e. as shown in Figures 5.4*d* to 5.4*f*). There is some critical departure from isostasy beyond which the deformation develops as in Figures 5.4*a* to 5.4*c* and below which the mode of deformation is the same as without the arc (Figures 5.4*d* to 5.4*f*). The isostatic state of the arc in nature is determined by many factors, such as the age of the arc (how long the plate has been under arc load) and its temperature regime, which in the arc area is characterized by the elevated temperature [*Furukawa*, 1993]. Thus the question which case in Figure 5.4 is more realistic is not obvious and must be discussed with regard to the particular arc considered. For example, the Luzon arc experiences subsidence under strong compression as it approaches the Taiwan collision zone (see Section 5.5), while the Kuril arc undergoes uplift (see next section). These data can mean that the arc in the Kuril zone is isostatically compensated and in the Luzon zone is not. However other explanations are also possible. For example, uplift of the Kuril arc can be caused by the lithospheric failure in the back arc area which might be thinner than the lithosphere directly in the arc. Failure is preceded by buckling of the back arc lithosphere which is accompanied by an uplift of the area between

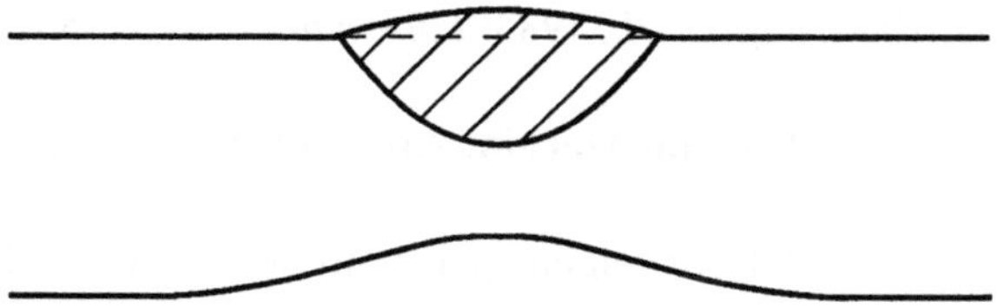

Figure 5.5. Model of the thinned plate with complete isostatic compensation of the volcanic arc .

this lithosphere and a thicker lithosphere of the arc, see Figure 5.2 (this effect is similar to experiments 3.3 and 3.4, Figures 3.3 and 3.5). On the other hand, the subsidence of the Luzon arc may not be solely caused by the weight of the arc (which might be isostatically compensated) but by the weight of the thick sedimentary layer accumulated in the fore arc basin, the Luzon Trough (see Sections 5.5.3 and 5.5.4).

5.4. Deformation of the Overriding Plate in the Kuril Subduction Zone

The Kuril arc is an example of a subduction zone with compressional tectonics in the back arc area recognized from both seismological and geological data [*Zonenshain and Savostin*, 1979]. According to geological data compression replaced tension and back arc spreading in the Okhotsk Sea (Kuril Trough) during Pliocene times, though the causes for such a change remain unclear. Seismicity in the rear of the Kuril subduction zone is concentrated along the south-eastern margin of the Kuril Trough (Figure 5.6). The earthquake hypocenters do not exceed 40-50 km in depth in this area and form a focal zone dipping beneath the arc (see inset in Figure 5.6). Both focal mechanisms for back arc earthquakes and aftershocks distribution indicate thrust faulting in the focal zone at the foot of the Greater Kuril arc [*Baranov and Lobkovsky*, 1980]. Sedimentary cover in the same zone has been found to contain anticlinal folds. Their geometry is consistent with fault movements in the focal zone inferred from seismological data [*Baranov and Lobkovsky*, 1980]. Finally, as was mentioned above, the Kuril arc has undergone stable uplift since the Quaternary period [*Fedorchenko and Piskunov*, 1977]. This picture resembles what was obtained in experiment 5.1 (Figure 5.2) and corresponds to an initial stage of the Kuril Trough closing (say, the stage in Figure 5.2*e*). A similar pattern appears in the Japan zone where initiation of subduction of the Japan Sea lithosphere under the western margin of the arc is also supposed by a number of authors [*Huzita*, 1980; *Nakamura*, 1983; *Tada*, 1984; *Tamaki*, 1984; *Sugi et al.*, 1983; *Lallemand et al.*, 1985].

5.5. Subduction of the Continental Margin: Arc-continent Collision in Taiwan

Taiwan is one of the most convenient places to study the subduction regime and deformation of the overriding plate under strong horizontal compression. Such a compression occurs in this area due to subduction of the Chinese continental margin under the Luzon arc. The overriding Philippine Sea Plate and especially the Luzon arc undergo intensive deformation and failure resulting in the formation of a new plate boundary

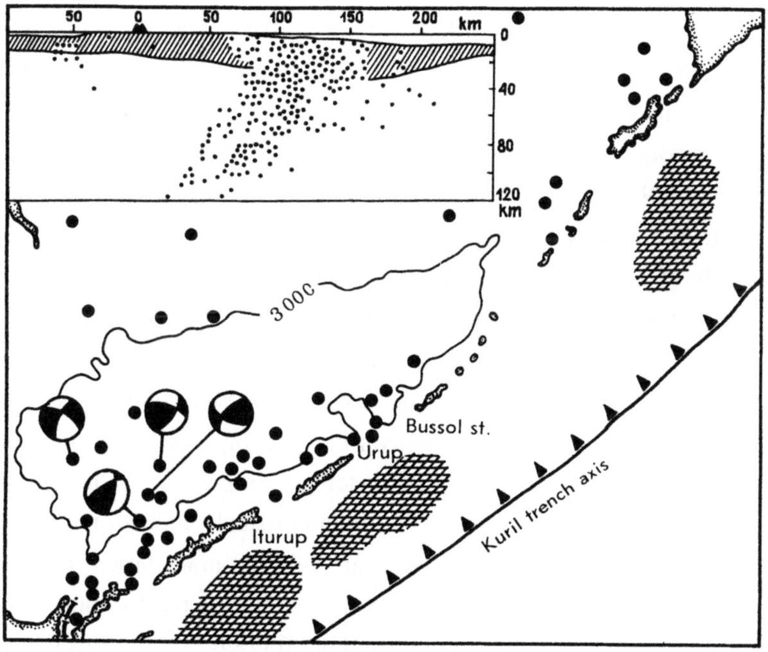

Figure 5.6. Shallow seismicity behind the Kuril arc (M ≥ 4) (after [*Baranov and Lobkovsky*, 1980]). Shaded areas are the regions of the maximum release of seismic energy.

offshore of north-eastern Taiwan [*Shemenda et al.*, 1992*a*,*b*, 1993]. In the present section we investigate this situation on the basis of relevant model experiments.

5.5.1. Geodynamic Setting of the Taiwan Area

Convergence of the Philippine Sea and Eurasian Plates in the Taiwan area at a rate of 7 cm/yr [*Seno*, 1977] is accommodated along the opposing Ryukyu and Luzon subduction zones and directly along Taiwan [*Seno*, 1977; *Biq*, 1972], Figure 5.7. Taiwan represents the north-western continuation of the Luzon subduction zone and formed only a few million years ago as a result of collision of this zone with the Asian continental margin [*Biq*, 1972; *Chai*, 1972; *Murphy*, 1973; *Karig*, 1973; *Wu*, 1978; *Suppe*, 1981; *Chi et al.*, 1981; *Barrier and Angelier*, 1986; *Ho*, 1988; *Teng*, 1990]. Due to the oblique orientation of the Luzon arc with respect to the continental margin, collision in Taiwan is propagating southward and incorporates ever new portions of the Luzon arc [*Page and Suppe*, 1981; *Lewis and Hayes*, 1983; *Suppe*, 1984]. The collision is thus more advanced in northern Taiwan and is in its initial stage to the south of Taiwan.

The main body of this collisional orogen comprises deformed accreted sediments of the Chinese continental margin [*Chang and Chi*, 1983; *Ho*, 1986; *Teng*, 1990]. The mechanism for accretion of the sedimentary layer seems to resemble that for well investigated accretionary prisms in "normal" subduction zones within deep sea trenches [*Karig and Sharman*, 1975]. The mechanism consists of successive underthrusting of ever new wedges of sedimentary material under the sediments that have already been accreted. The difference between the accretion in classical subduction zones and in the collision zone roughly is only in the spatial scale which is much larger in the latter case. During the collision a huge accretionary prism approaches the fore arc basin and the volcanic arc which both undergo contraction and uplift [*Teng*, 1990]. This general scheme seems to be reasonable and does not meet any serious objection. The problem is the particular mechanism for the deformation and failure of the overriding Philippine Sea Plate responsible for the shortening and rising of both the fore arc and arc areas.

Convergence between the plates during the relatively early stages of collision corresponding to the southern margin of Taiwan is accommodated in a subduction of the Asian Plate under the Luzon arc. In the northern Taiwan this process seems to have almost ceased. There is neither evidence for continuing accretion of the sediments of the continental margin in this area nor for considerable active shortening within the orogen which would correspond with the fast convergence of the Asian and Philippine Sea Plates.

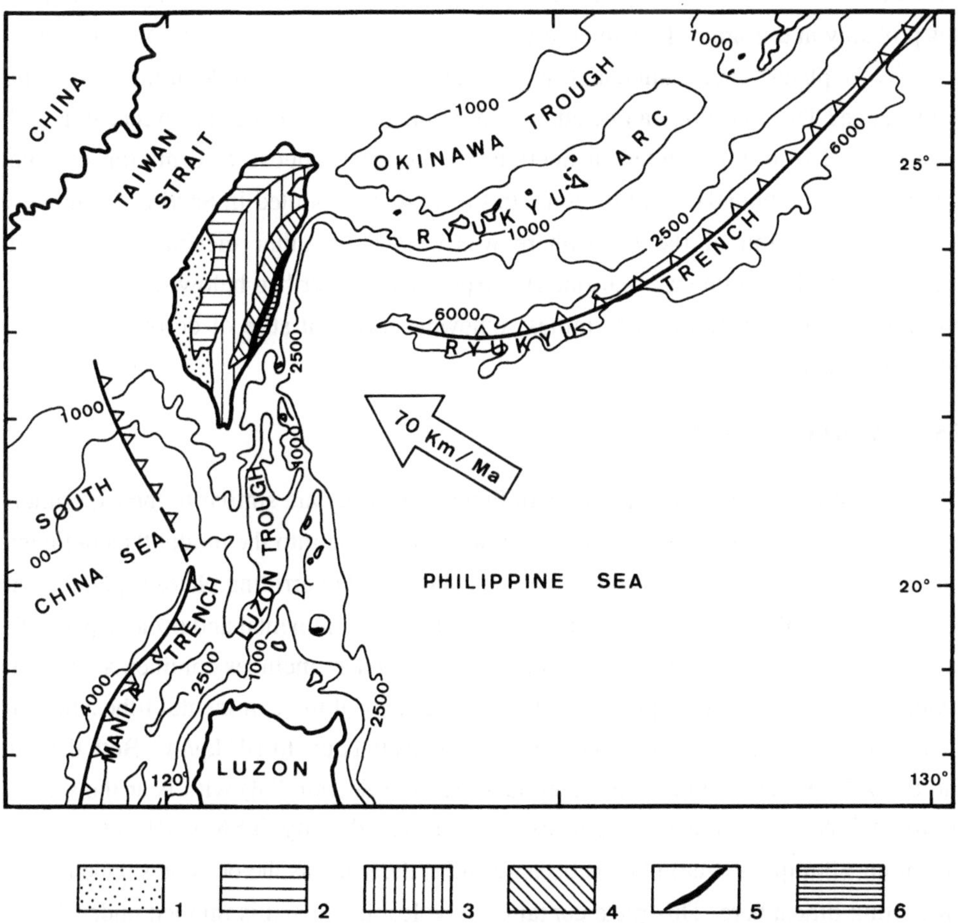

Figure 5.7. Geodynamic setting of the Luzon-Taiwan region [after *Ho,* 1986] (north-western continuation of the Manila trench (deformation front) is from *Lundberg et al.* [1991]).
1, Coastal Plain; 2, Western Foothills; 3 and 4, Central Range (4, Tananao belt); 5, Longitudinal Valley; 6, Coastal Range.

So, where this convergence is accommodated? It must occur where the majority of seismic activity associated with the collision takes place. The map in Figure 5.8 shows that seismicity is concentrated along the eastern margin of Taiwan, being most intensive in the offshore area (seismicity to north-east of Taiwan, seen on the map, is associated with both the subduction under the Ryukyu arc and the opening of the Okinawa Trough). Therefore, it is probably in this area where the most intensive deformation develops and convergence between the plates is accommodated. What is the mechanism and style of the deformation in this zone? This question can be addressed by the physical modeling. According to the experimental results presented in Section 5.3 one can suppose a formation of a new convergent plate boundary in this zone. In the models presented in Section 5.3 however strong compression of the lithosphere was caused by other mechanisms then the subduction of the continental margin. The experiments described below show that such a difference is important and determines somewhat different stile of deformation of the overriding plate.

5.5.2. Experimental Setting

Scheme of the experiments is shown in Figure 5.9. They are similar to the experiments described above, the difference is that now we incorporate into the model the continental and arc crust. The lithosphere consists of subducting and overriding plates separated with an inclined cut. The subducting (Asian) plate comprises continental and oceanic parts. The oceanic plate is one-layer while the continental lithosphere includes continental crust and mantle layer. Oceanic lithosphere and the mantle layer of the continental lithosphere are made of the same material (compound N1) with density equal to 10^3 kg/m^3. Both the crust and the volcanic arc are made of compound N2 (see Section 2.4) with near the same as compound N1 mechanical properties but a lower density 0.86×10^3 kg/m^3. The overriding (Philippine Sea) plate is oceanic and is the same as the oceanic segment of the subducting lithosphere. This plate contains the volcanic arc and is thinned under the arc. We used the same technique for the preparation of the crust and the arc as that for the aseismic ridge in experiment 5.1.

Similarity criteria comprise the same conditions as for the previous experiments (2.1) plus criteria imposing limitations on the parameters of the crust and the volcanic arc. The whole set of similarity criteria is the following

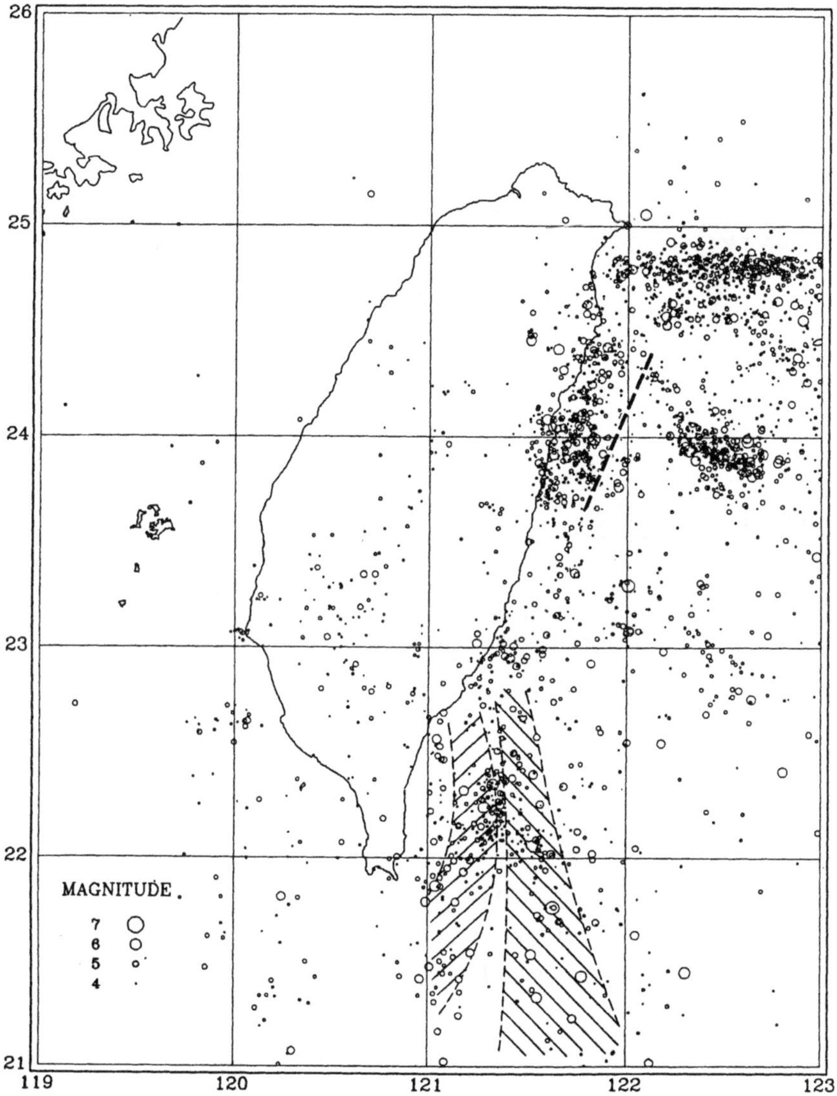

Figure 5.8. Seismicity map for earthquakes $M_d \geq 4$ located during 1973-1993 by the TTSN thick dashed line shows the axis of the gravity low which is dotted in Figure 5.17; dashed areas roughly show the Huatung Ridge (to the west) and the Luzon arc (to the east) according to *Huang et al.* [1992], see also Figure 5.15.

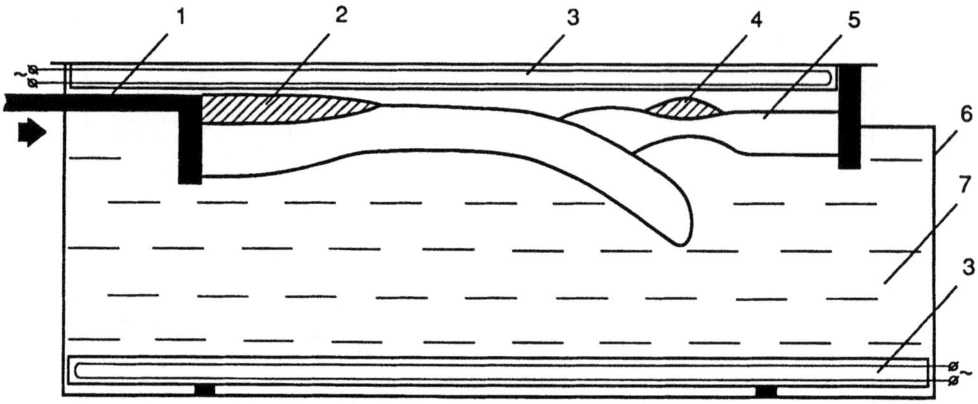

Figure 5.9. Scheme of the model experiments and the experimental installation
1, piston; 2, subducting Asian Plate; 3, heaters; 4 volcanic (Luzon) arc; 5, overriding (Philippine Sea) lithosphere; 6, tank; 7, asthenosphere (dashed are the continental crust and crust of the volcanic arc).

$$\sigma_s/(\rho_l g H_l) = \text{const};\ \sigma_s/\sigma_c = \text{const};\ \sigma_s/\sigma_v = \text{const};\ E/(\rho_l g H_l) = \text{const};$$
$$\rho_l/\rho_a = \text{const};\ \rho_l/\rho_v = \text{const};\ \rho_l/\rho_c = \text{const};\ H_l/H_c = \text{const};\ H_l/H_v = \text{const}; \quad (5.1)$$
$$H_m/H_c = \text{const};\ d/H_v = \text{const};\ Vt/H_l = \text{const};$$

where σ_s, E, ρ_l as previously are the yield limit for normal loading, Young's modulus and the average density, respectively, for both the oceanic lithosphere and the mantle layer of the continental lithosphere; ρ_c and σ_c are the density and the yield limit for the crust, respectively; ρ_a is the asthenospheric density; H_l, H_c, H_m, and H_v, are the thicknesses of the oceanic lithosphere, the continental crust, the mantle layer of the continental lithosphere, and the crust in the volcanic arc, respectively; d, ρ_v and σ_v are the width, the density, and the strength of the volcanic arc; V is the rate of subduction; t is the time; g is the acceleration of gravity.

The following parameter values for the prototype (nature) and the model satisfying criteria (5.1) are assumed. For the original (prototype): $\sigma_s^o = \sigma_c^o = \sigma_v^o = 2.1 \times 10^8$ Pa; $E^o = 1.5 \times 10^{10}$ Pa; $\rho_l^o = \rho_a^o = 3.3 \times 10^3$ kg/m^3; $\rho_c^o = \rho_v^o = 2.84 \times 10^3$ kg/m^3; $H_l^o = 5 \times 10^4$ m; $H_c^o = 3.6 \times 10^4$ m; $H_m^o = 5.2 \times 10^4$ m; $H_v^o = 1.4 \times 10^4$ m; $d^o = 5.2 \times 10^4$ m; $g^o = 10$ m/s^{-2}. For the model: $\sigma_s^m = \sigma_c^m = \sigma_v^m = 28$ Pa; $E^m = 2 \times 10^3$ Pa; $\rho_l^m = \rho_a^m = 1 \times 10^3$ kg/m^3; $\rho_c^m = \rho_v^m = 0.86 \times 10^3$ kg/m^3; $H_l^m = 2.2 \times 10^{-2}$ m; $H_c^m = 1.6 \times 10^{-2}$ m; $H_m^m = 2.3 \times 10^{-2}$ m; $H_v^m = 0.6 \times 10^{-2}$ m; $d^m = 2.3 \times 10^{-2}$ m; $g^m = 10$ m/s^{-2}. Superscripts "o" and "m" denote the original and the model, respectively.

5.5.3. Experimental Results

A total of 20 experiments were carried out under different conditions. Below we present the results of three experiments conducted with the parameters indicated above. The experiments differ only in the thickness of the lithosphere under the arc and the friction between the plates. Special lubrication is used to reduce the friction.

Experiment 5.2. Friction between the plates is very small. The initial stage of the experiment (Figure 5.10) is similar to experiments 3.8 and 4.1 (Figures 3.9 and 4.1). The subducting plate bends down forming the trench (Figure 5.10*b*). The most frontal parts of the overriding plate rise while the segment of this plate to the right (to the east) undergoes subsidence. The uplift within the overriding wedge migrates to the east during initial subduction and widens causing uplift of the segment of the overriding wedge that initially subsided (Figure 5.10*c*). The frontal parts of the wedge descend into the trench after the rise has passed to the east. The picture then stabilises (the subduction becomes quasi steady state in similar experiments with a longer oceanic segment of the subducting plate) until the continental margin with light crust starts to subduct. Both the further rapid uplift of the whole overriding wedge and the increase in horizontal compressive stress occur at this stage. The area of the volcanic arc then undergoes flexural subsidence and deformation similar to those shown in Figures 5.4*a* and 5.4*b*. Concentration of deformation in this area leads to thickening of the lithosphere and the formation of two major inclined zones of localization of thrust (shear in cross section) deformation dipping under the arc from both sides (Figures 5.10*d* and 5.10*e*, see also Figure 5.4*c*). Complete failure of the plate then occurs as expected along the zone dipping to the west (Figure 5.10*f*) resulting in underthrusting (subduction) of the Philippine Sea Plate under the arc. The subduction of

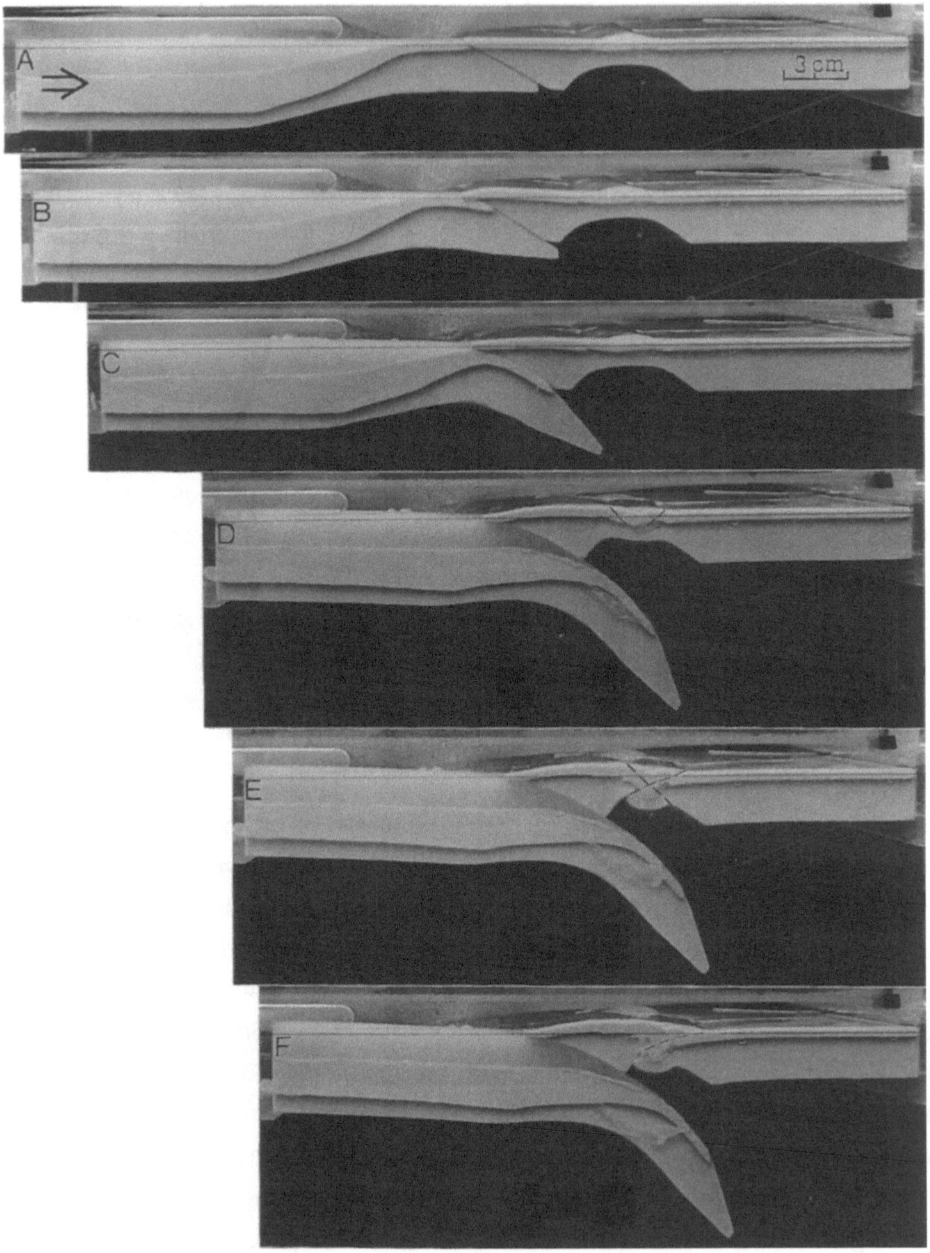

Figure 5.10. Result of experiment 5.2 (friction between the plates is very low).

the continental margin slows down.

Experiment 5.3, Figure 5.11. Friction between the plates is once again low. The thickness of the lithosphere under the arc is 15% less than in experiment 5.2. The initial stages of experiment 5.3 (Figures 5.11*a* to 5.11*c*) are similar to those in the previous experiment. However, plastic shortening of the arc area and formation of two major thrust zones begin earlier. Complete failure of the lithosphere then occurs along a zone dipping under the arc from the west, Figures 5.11*f* to 5.11*h* (unlike in the previous experiment) producing underthrusting of the frontal wedge of the Philippine Sea Plate under the arc. This process, however, proved to be unstable. We conducted several experiments under the conditions of experiment 5.3 and obtained two possible scenarios for the continuation of the process after the formation of the fault dipping under the arc from the west. First, the underthrusting of the former frontal wedge of the Philippine Sea Plate continues up to the complete subduction of this wedge into the mantle (as described in Section 5.3, Figure 5.3*e*). Second, the underthrusting stops at a certain stage, resulting in reactivation of subduction of the continental margin, followed by lithospheric failure along the other zone dipping under the arc from the east as seen in Figure 5.12. It was observed that the process depends on the length of the continental margin (continental crust) that had been subducted before lithospheric failure. The length in turn depends on the friction between the plates which can not be precisely adjusted and varies slightly from one experiment to another. The higher the friction, the stronger the compression and the earlier the lithosphere fails. The earlier the underthrusting under the arc begins, the higher the probability for the frontal wedge of the Philippine Sea Plate to be completely subducted.

In some experiments we intentionally increased the friction between the plates. This led to initial failure of the lithosphere along a zone dipping under the arc to the east even when the plate under the arc was thicker. Figure 5.12 shows the results of experiment 5.4 conducted under the same conditions as experiment 5.2, only the friction was higher. The plate fails in this experiment first along the fault dipping to the east resulting in some underthrusting and intensive plastic deformation of the wedge overriding the continental margin. Due to the deformation, the upper part of the initially eastward dipping fault (dotted line in Figure 5.12) changes the dip on the opposite and finally the underthrusting develops under the arc from the east along the dashed line in Figure 5.12.

Let us consider the deformation mechanism of the models. Figure 5.13*a* shows the configuration of the plates before deformation. The plates are in isostatic equilibrium and the surface of the overriding plate is horizontal except in the area of the volcanic arc where

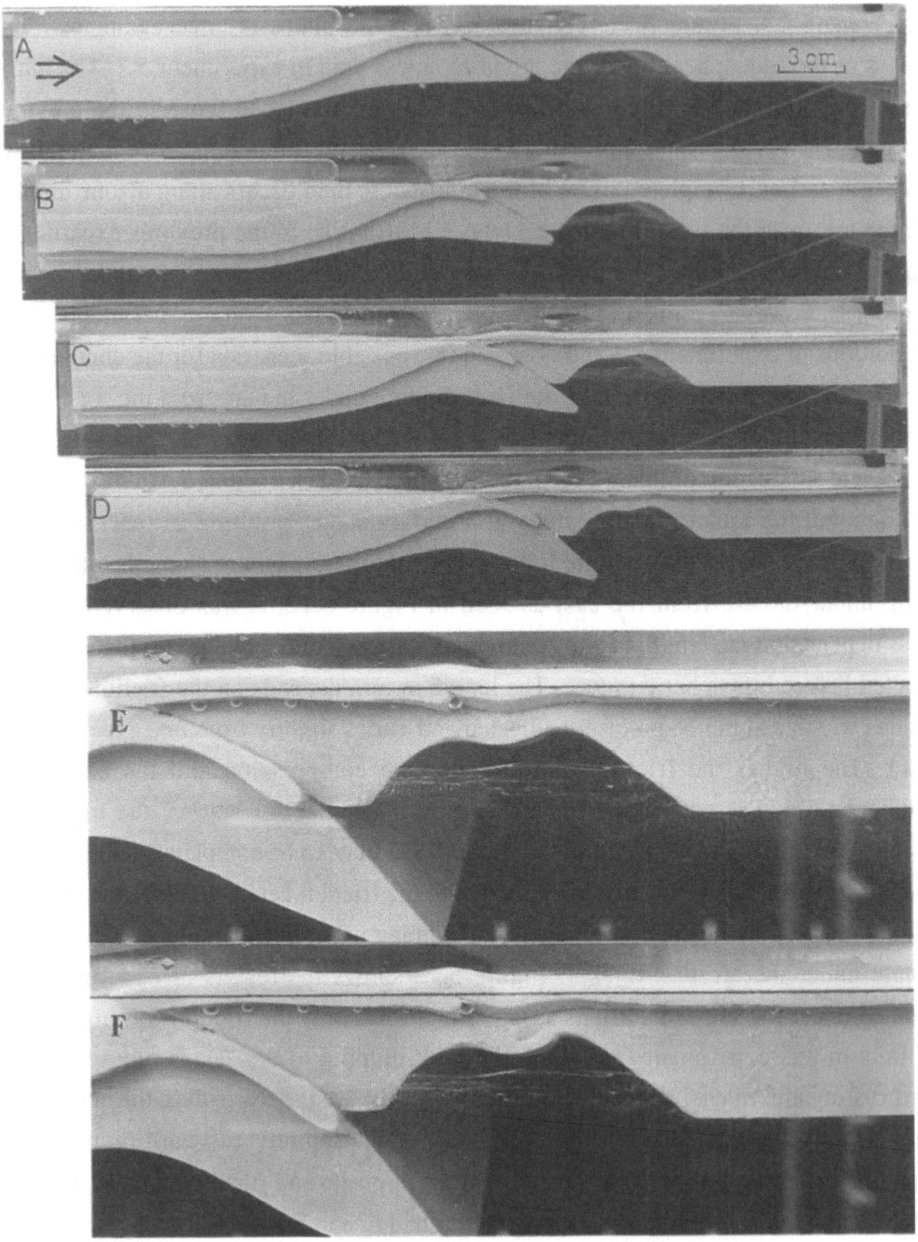

Figure 5.11 (see figure caption on the next page)

Figure 5.11. Result of experiment 5.3 (friction is very low; thickness of the plate beneath the arc is 15% less than in experiment 5.2).

the lithosphere is bent down under the weight of the arc. Compression exerted by the piston forces the Asian Plate to subduct. The pressure between the plates increases during this process causing the subducting plate to bend down and the overriding plate to rise up. At the beginning of subduction, the superhydrostatic pressure is maximal at a shallow depth just under the elevated part of the overthrusting wedge (Figures 5.10*b*, 5.11*b* and 5.11*c*). The leading wedge of the subducting plate experiences bending and clockwise (in cross section) rotation under the pressure, pulling down the above segments of the overthrusting wedge due to hydrostatic suction between the plates (see Section 4.3). As subduction matures, the zone of maximal pressure is shifted down producing an elevation of the whole overthrusting wedge (except the most frontal parts) and formation of the frontal arc as seen in Figure 5.10*c* (see also Figures 4.2 and 4.9). The horizontal component of the superhydrostatic pressure force F_p (Figure 5.13*b*) produces horizontal compression of the lithosphere. At the stage of subduction of the oceanic segment of the

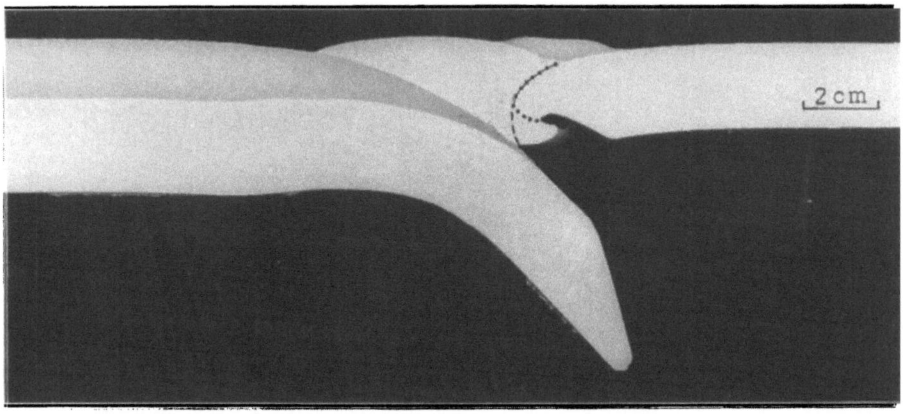

Figure 5.12. Result of experiment 5.4 (the photo of the model section was made after freezing the deformed "lithosphere"). Thickness of the plate under the arc is the same as in experiment 5.2, the friction between the plates is higher. Dotted line shows strongly deformed one of the major thrust zones which initially dipped to the right. Final failure of the plate occurred along the dashed line and the dots linked by the solid lines.

Asian Plate, the nonhydrostatic compressive stress σ_h in the plate segment of thickness h reaches a value near $0.3H_l\sigma_s/h$ (see Section 3.2). Subduction of the continental margin with light crust results in an increase in nonhydrostatic contact pressure σ_n (see Figure 4.9), the force F_p, and consequently, the lithospheric compression, and the nonisostatic elevation of the frontal arc. High compression leads to flexural buckling of the overriding plate in the thinned zone (the volcanic arc area) and its subsidence (similar to Figure 5.4*b*) At a specific stage, σ_h reaches the yield limit of the lithosphere in the arc area resulting in its plastic deformation and shortening. Roughly, the deformation develops along two slip

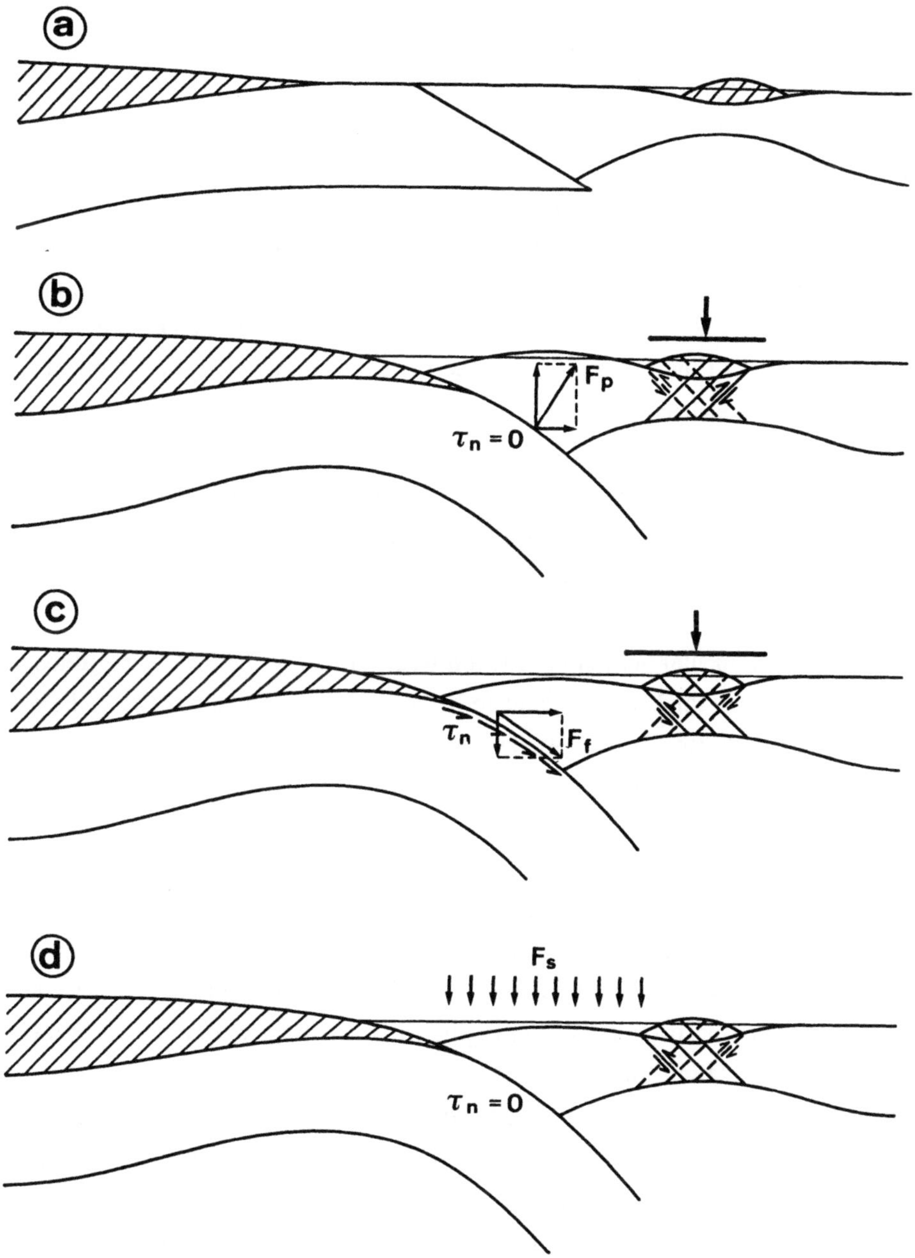

Figure 5.13. Scheme of the model deformation (see text for explanation).

line families dipping at an angle of 45° in opposite directions (Figure 5.13*b*). The deformation then concentrates along two oppositely dipping narrow zones located near the boundaries between the plastic zone and a rigid plate (this final stage of deformation localization is also shown in Figures 5.4*c* and 3.4*b*). Complete failure of the plate occurs along the zone dipping to the west (solid line in Figure 5.13*b*).

In experiment 5.3 (Figure 5.11), lithospheric failure begins during the early stages of subduction when the overriding wedge is pulled down by the suction force. Therefore, the complete failure of the plate occurs along the zone dipping to the east. A similar effect is produced by high friction τ_n between the plates (Figure 5.13*c*, see also Figure 5.3*d*). The vertical component of the friction force F_f acting against the vertical component of F_p drags the overriding wedge down and makes the dip direction of lithospheric failure to the east more favourable, Figure 5.13*c* (the result depends, of course, on how large the friction is comparable to the superhydrostatic pressure between the plates). Compressive stress σ_h in this case is higher due to the horizontal components of both the pressure and the friction forces. This is the reason why failure of the plate occurred earlier in experiment 5.4.

The stage of continental margin subduction when failure of the overriding lithosphere occurs depends on the thickness of the plate in the arc and the friction between the plates. The thicker the plate (the less its weakening) and the lower the friction, the later the compressional stress reaches the yield limit. The longer the segment of the continental margin and the thicker the continental crust that have been subducted, the higher both the elevation of the overriding wedge and the probability of immediate failure of the lithosphere along the fault dipping to the west (as in Figure 5.13*b*). If the lithospheric yield limit is reached too late, then the plate fails along this fault anyway, and it does not matter what the friction value is (because the force F_p and its vertical component sooner or later become larger than the friction force dragging the overriding wedge down). However, given that the plate failure first occurs along a zone dipping to the east (Figure 5.13*c*), underthrusting of the frontal wedge of the Philippine Sea Plate along this zone can not be stable unless the failure occurs at a very early stage of continental margin subduction (as in Figure 5.11). The wedge already takes a high position being overridden on the continental margin. The underthrusting along the east dipping zone causes again an increase in compression and then failure of the plate occurs along another zone as in Figure 5.12. In reality, a blockage of the underthrusting along the eastward dipping zone should occur even sooner because the underthrusting wedge (former frontal wedge of the overriding plate) usually has "sedimentary-crustal" composition with average density lower than the mantle density (in the experiments densities of the wedge and the mantle are equal).

Which of the two options (Figures 5.13*b* or 5.13*c*) for lithospheric failure is more realistic? An analysis of the balance of forces causing extension in the overriding plate in subduction zones with active back arc basins leads to the conclusion that friction τ_n between the plates in the subduction zone is typically very small (see Sections 4.4 and 4.19). Therefore, immediate failure of the lithosphere in the Luzon arc area along the fault dipping to the west seems to be more likely. However, one can imagine another situation where the original fault in the lithosphere dips to the east even at $\tau_n = 0$. This would occur if there is some force F_s distributed along the surface of the overriding wedge pushing the wedge down (Figure 5.13*d*). This force could be due to the weight of the sediments accumulated in both the accretionary prism and the fore arc basin. The Luzon subduction zone near the southern margin of Taiwan indeed has a huge accretionary prism [*Lundberg et al.*, 1991] and a large fore arc basin, the North Luzon Trough [*Huang et al.*, 1992]. Besides the interplate friction at the stage of collision can also grow. Therefore, the probability for the immediate failure of the Philippine Sea Plate along the eastward dipping zone still remains.

The modeling thus does not provide us with a definite unambiguous scenario of lithospheric deformation in the Luzon-Taiwan area. The process depends on many parameters and factors which are not well constrained, such as the friction between the plates, the value and distribution of the sedimentary load on the overriding wedge, the structure and density of the frontal wedge of the Philippine Sea Plate, the extent of the weakening and geometry at the base of the overriding plate in the arc area, the strength of the coupling between the arc and underlying plate, and any pre-existing lithospheric inhomogeneities. The experiments provide us only with ideas about the possible evolution of deformation. The resulting geodynamic model of the collision should be based on both these ideas and the observational evidences. Below we attempt to create such a model.

5.5.4. Evolutionary Model for the Collision in Taiwan

Figure 5.14*a* shows the Luzon subduction zone before its collision with Asia. Since collision in Taiwan is oblique, this picture can correspond to both the current situation in the Luzon arc to the south of Taiwan and in northern Taiwan a few million years ago. The Luzon subduction zone is believed to have originated within the oceanic basin. Therefore,

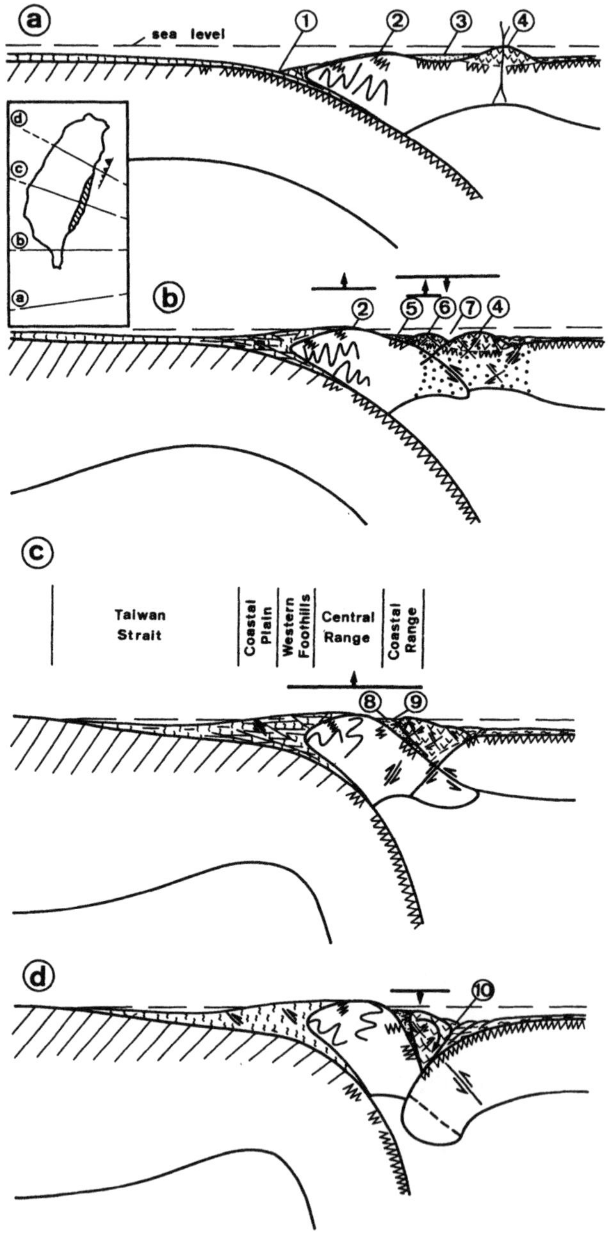

Figure 5.14a-d. An idealised evolutionary model for the Luzon-Taiwan region (see figure caption on the next page).

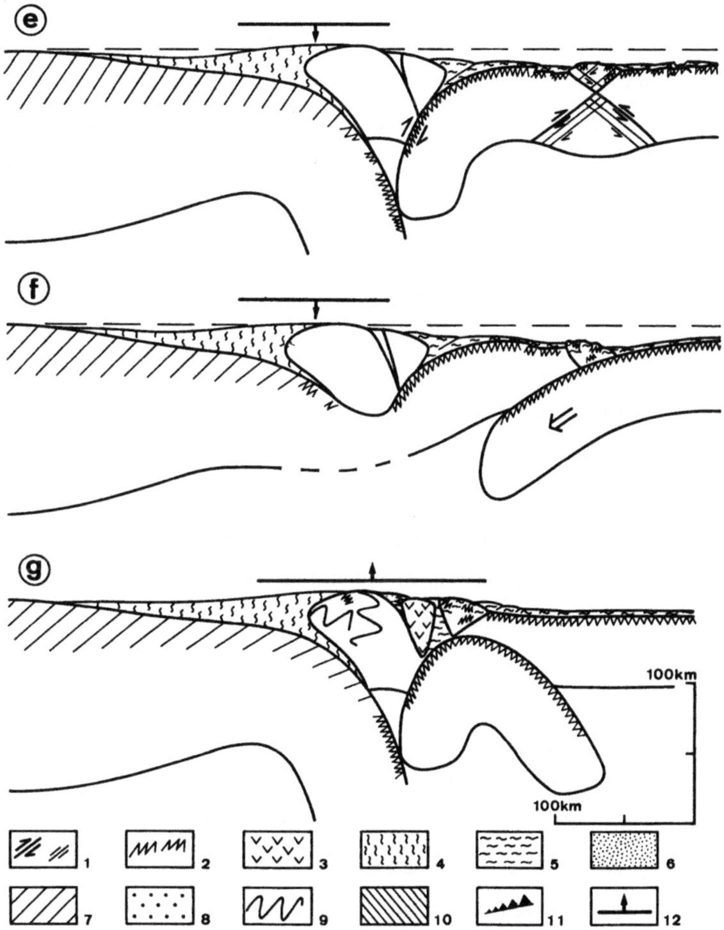

Figures 5.14e-g. A possible continuation of the process shown in Figures 5.14*a-c* (see text for more explanation)

1, faults; 2, oceanic crust; 3, Neogene volcanics; 4, continental margin sediments; 5, Philippine sea sediments; 6, fore arc basin sediments; 7, continental crust; 8, zone of plastic shortening of the Philippine Sea Plate; 9, overthrusting wedge of the Philippine Sea Plate formed during previous evolution of the Asian margin (see text); 10, Coastal Range (inset); 11, supposed incipient southward propagating subduction zone (inset); 12, direction of vertical movements.
① Luzon Trench; ② Hengchun Ridge; ③ North Luzon Trough; ④ Luzon arc; ⑤ Southern Longitudinal Trough; ⑥ Huatung Ridge; ⑦ Taitung Trough; ⑧ Longitudinal Valley; ⑨ Longitudinal Valley Fault; ⑩ Axis of plate convergence off north-eastern Taiwan.

the volcanic arc and part of the fore arc area is probably underlain by oceanic crust while the overriding wedge located to the west has an unclear origin and composition inherited from the previous evolution of the Chinese continental margin (see Section 5.5.6). Impingement of the Philippine Sea Plate on the continental margin results in off-scraping of a thick sedimentary layer of the margin, elevation of the overriding block (the Hengchun Ridge ②, an increase in the lithospheric compression, and the complete cessation of volcanism. Flexural subsidence, plastic deformation (shortening) of the lithosphere, and intensive thrusting begin in the arc - fore arc area (dotted in Figure 5.14*b*) when the compressive stress reaches the lithospheric yield limit. The arc (the volcanics) itself can be involved into intensive thrusting or not depending on coupling between the arc and underlined lithosphere. At first, the deformation is concentrated near the western margin of the arc producing (squeezing up) a fractured high, ⑥, the Huatung Ridge, Figure 5.14*b* and Figure 5.15 (a corresponding high is also seen in the experimental pictures in Figures 5.11*e* to 5.11*h*). The ridge is composed of strongly deformed fore arc sediments, the western margin of the volcanic arc, and the underlying rocks of oceanic crust. Thrusting is concentrated later along two major zones dipping under the arc from the west and the east with more intensive deformation along the former zone which is the proto-Longitudinal Valley Fault. Subduction of the Asian Plate and uplift of the frontal wedge, the Hengchun Ridge (which further to the north becomes in our model a part of the Central Range) gradually slows down. The volcanic arc (block) and underlying Philippine Sea lithosphere overthrust the frontal wedge along with deformed fore arc complex. The rising western edge of the volcanic block together with part of the Huatung Ridge form the Coastal Range with the Longitudinal Valley, ⑧, in front of it (Figure 5.14*c*). Overthrusting along the Longitudinal Valley Fault, ⑨, then ceases causing a further increase in the compressive stress. Thrusting is then concentrated along the zone dipping from the eastern margin of the arc to the west. Failure of the lithosphere occurring along this zone decreases the compression and initiates westward underthrusting of the Philippine Sea Plate under the eastern margin of Taiwan (Figure 5.14*d*).

The further evolution depends on the interaction between the subducting Philippine Sea lithosphere and the Asian Plate subducted earlier. The presence of this subducted Asian Plate forces the Philippine Sea Plate to subduct steeply, which requires greater compression. (One can also suppose that steeply subducting plate experiences quasi stationary thrust deformation developing along the zone dipping to the east as shown in Figure 5.14*d*, see also Figures 3.17 and 3.22). If the subducted segment of the Asian

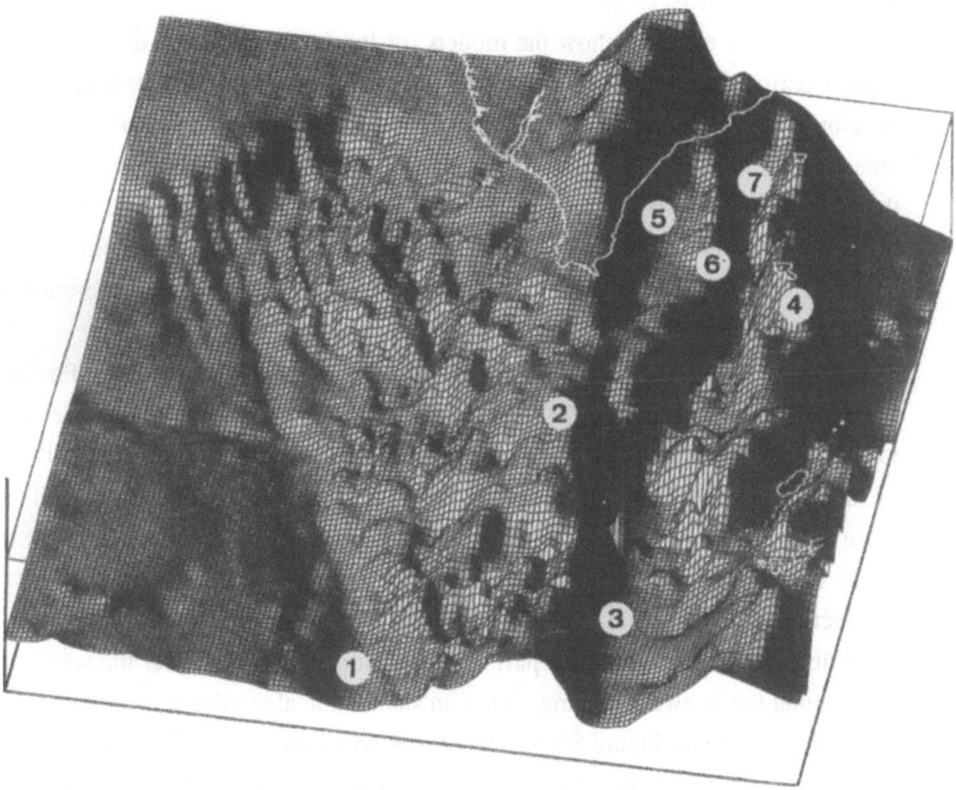

Figure 5.15. Physiographic diagram of the southern margin of Taiwan [from *Liu et al.*, 1992]
① Luzon Trench; ② Hengchun Ridge; ③ North Luzon Trough; ④ Luzon arc; ⑤ Southern Longitudinal Trough; ⑥ Huatung Ridge; ⑦ Taitung Trough.

Plate is sufficiently weakened by heating in the mantle, then the subduction under eastern Taiwan would be stable. If conversely, the resistance from the Asian Plate is large then collision of the subducted plates at depth would cause a further increase in compression and a new phase of deformation in both the orogen and the Philippine Sea Plate. The result may be the reactivation of the underthrusting of the Asian Plate under the collisional orogen (the continental subduction or obduction) or further reorganization of the plate boundaries to the east of Taiwan.

Figures 5.14*e* to 5.14*g* show the models for the future evolution of the collision. The models correspond to the continued deformation of the Philippine Sea Plate which could develop if the collision process were two dimensional. In fact, the situation in the Taiwan area (at least in its northern part) is essentially three dimensional (see Section 5.5.6) and therefore the models based on the results of 2-D experiments probably are not applicable to this region. We present them in this work only to illustrate a possible continuation of the arc-continent collision in principal which may be useful in the consideration of other simpler collision zones.

According to Figure 5.14*e* a blockage of the incipient subduction of the Philippine Sea Plate under the north-eastern Taiwan results in flexural buckling of this plate and then in localization of deformation at a distance of $\lambda/2$ (~100 to 150 km) to the east (this is the type of subduction zone jump illustrated in Figure 5.1). A new subduction is initiated at this place in one of two possible directions, to the west or to the east causing compressive stress to drop. If failure occur along the west dipping fault (Figure 5.14*f*) then the subduction will be stable. The Taiwan orogen will subside in this case to restore an isostatic equilibrium, with most part of the island submerging under the water. If failure occur along the eastward dipping fault then subduction along this zone will result in a new collision as shown in Figure 5.14*g*. The orogen will continue development and growing in this case. Such a scenario of evolution may result in the formation of an accretionary structure similar to Kamchatka [*Koltypin*, 1991].

5.5.5. Comparison of the Model with Observations

Uplift of the Huatung Ridge, ⑥, (Figures 5.14*b* and 5.15) in the early stages of collision due to intensive deformation and squeezing up of the upper lithospheric layers produces two troughs on both sides of the ridge. These can be identified as the Southern Longitudinal Valley Trough, ⑤, and the Taitung Trough, ⑦, (Figures 5.14*b* and 5.15). According to the model the southern tip of this ridge roughly marks the location where the compressive stresses reach the lithospheric yield limit in the arc - fore arc area initiating its plastic shortening. As seen in Figure 5.8, the seismic activity increases from this location to the north. A seismicity cross-section in Figure 5.16*e* shows the subducted Asian Plate, the Benioff zone (it is seen more clearly for $M_L \geq 3.5$) and the triangular-shaped zone of high seismicity which may be treated as a zone of localization of plastic deformation. During continued collision the former fore arc basin is closed due to the underthrusting along the east dipping zone, the Longitudinal Valley Fault, leaving a narrow suture zone, the Longitudinal Valley, ⑧, (Figure 5.14*c*). The Huatung Ridge underthrusts and is accreted to the western margin of the volcanic arc. The Taitung Trough is thus closed.

Uplift of the frontal wedge of the Philippine Sea Plate overriding the continental margin has both isostatic and nonisostatic components. The isostatic component is due to the light crustal material underthrusting the wedge. The nonisostatic component of relief is due to an increase in the nonhydrostatic pressure σ_n exerted by the thick and comparatively strong Asian Plate on the wedge. Both components, though mainly the nonisostatic one, cause the free air gravity anomaly to grow from near zero (or even negative values [*Bowin et. al.*, 1978]) over the Hengchun Ridge off Southern Taiwan to ~ 300 mGal further to the north in the Central Range (Figure 5.17).

The volcanic block rises up while overriding the Central Range along the Longitudinal Valley Fault (Figure 5.14*c*). Its western edge rises above sea-level and makes up the Coastal Range. Both the Central Range and the Coastal Range are thus rising, though the latter rises faster because it overrides the Central Range. Such vertical movements are corroborated by the geological and geodetic data [*Hsu*, 1976; *Yu and Tsai*, 1982; *Barrier*, 1985; *Liu and Yu*, 1990; *Yu et al.*, 1990].

Stratigraphic records in the Coastal Range show that the arc - fore arc area underwent rapid subsidence just before and during the initial stages of the collision [*Chang and Chi*, 1983; *Dorsey*, 1992]. According to the model such subsidence can be first due to the flexural bending of this area (before the localization of deformation in the arc lithosphere) and then due to the bending down of only the fore arc basin during its underthrusting along the Longitudinal Valley Fault (Figure 5.14*b*).

The eastward subduction of the Asian Plate practically ceases at stage *c* (Figure 5.14). The sediments intensively eroded from both the growing orogen and the continent cover the accretionary prism forming the Western Foothills and the Coastal Plain. The deformation front in the continental margin sediments during subduction is supposed to be somewhere in the Taiwan Strait, but is presently covered by undeformed sediments. The

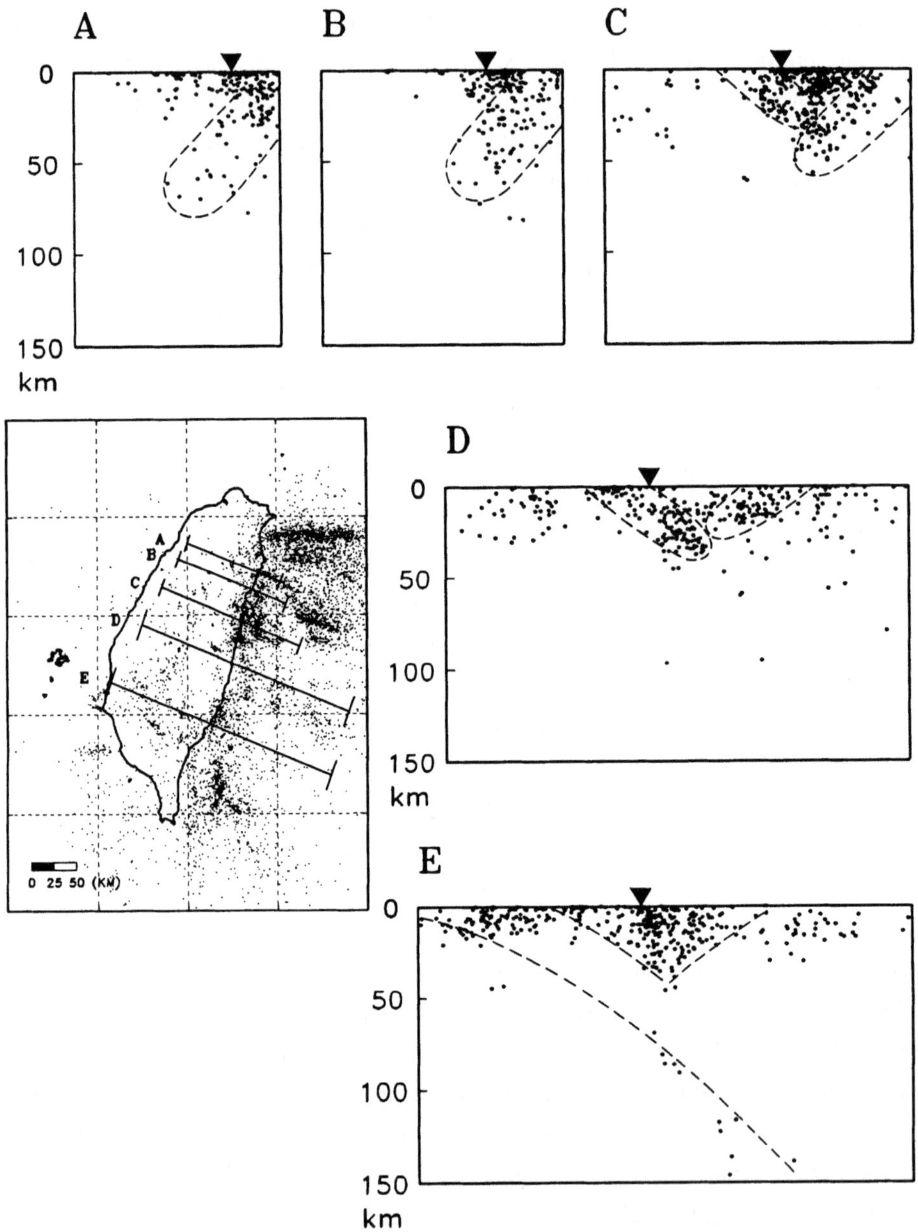

Figure 5.16. Seismicity cross-sections for the earthquakes $M_L \geq 4$ located during 1973-1993 by the TTSN
triangle shows the position of the eastern coastal line

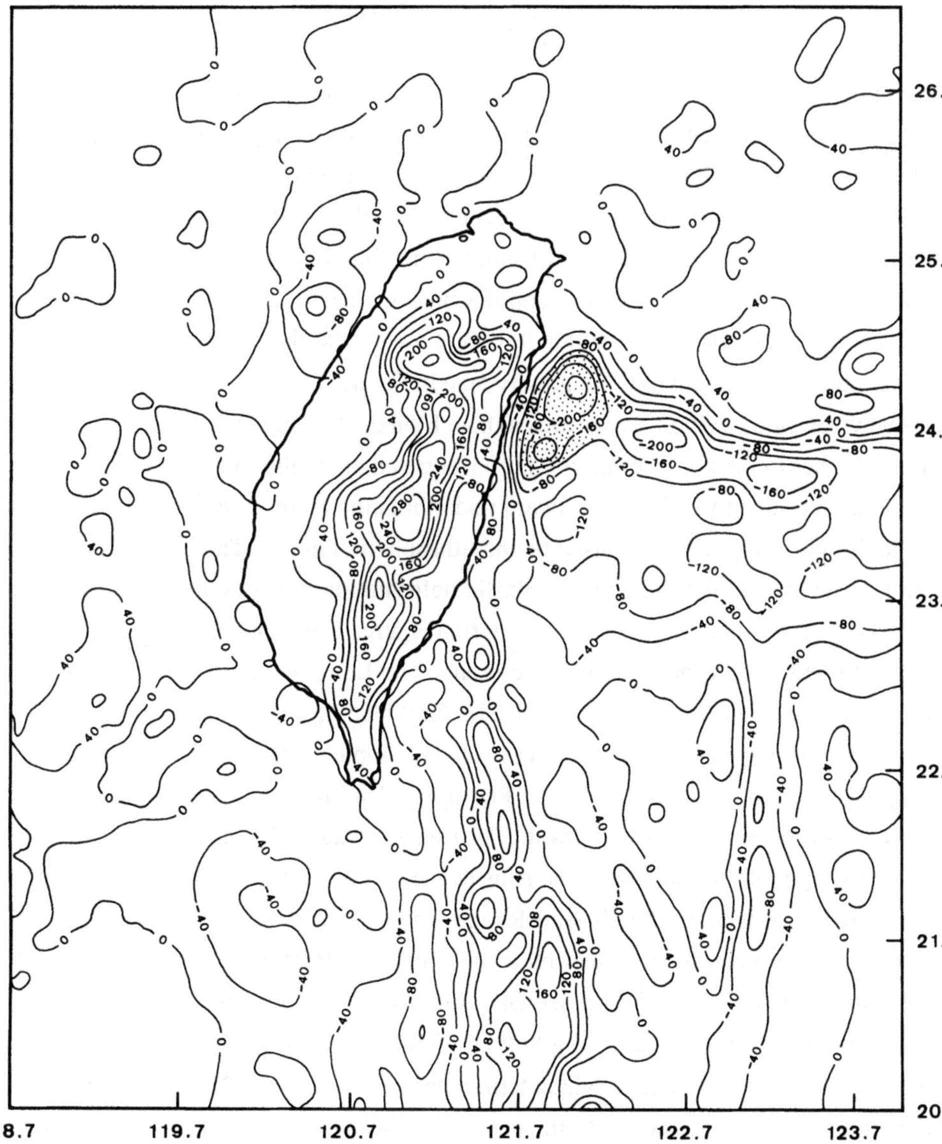

Figure 5.17. Free air gravity anomalies map (created by G. C. Tsuei combining the satellite altimetry data (Geos-3, Seasat and Geosat) and a map by *Yeh and Yen* [1991] for Taiwan).

front of the sediment deformation is observed up to a latitude of 22.2°N (Figure 5.15). Further to the north it may be covered by undeformed sediments. This latitude might thus mark the northern boundary of active eastward subduction of the Asian Plate. Further to the north, convergence between the Asian and Philippine Sea Plates should be accommodated, according to the model, mostly in the Luzon arc - fore arc area due to both diffusive plastic shortening of this area and the underthrusting along the Longitudinal Valley Fault. Blocking of the underthrusting along this fault causes the compressive stress in the lithosphere to reach its maximum. A complete failure of the Philippine Sea Plate then develops along the westward dipping fault zone. Maximum compression produces a maximum nonisostatic uplift of the Central Range and hence, a maximum free air gravity anomaly. The seismicity cross-section in Figure 5.16*d* near the location where Δg reaches its maximum of 300 mGal (see the gravity map in Figure 5.17), shows that the lithosphere is actively deforming within a wide area although subduction of the Asian Plate has already ceased. Further to the north seismicity is concentrated along the eastern margin of Taiwan (see Figure 5.8) due to both the westward underthrusting there of the Philippine Sea Plate and the drop in compression (caused by the lithospheric failure). The seismicity profile in Figure 5.16*c* shows that the zone of plastic deformation narrows and the Longitudinal Valley Fault is still active. Further to the north (Figures 5.16*b* and 5.16*a*) the deformation is concentrated within the ever narrower zone. The Longitudinal Valley Fault does not exhibit itself while the underthrusting (subducting) leading segment of the Philippine Sea Plate can be fairly well recognized in the westward dipping swarm of earthquakes.

Supposed westward subduction of the Philippine Sea Plate must produce the trench, ⑩, (Figure 5.14*d*) off north-eastern Taiwan. The trench is caused by the down bending of the underthrusting plate which causes the subsidence and down tilting of the frontal parts of the overthrusting wedge, the Coastal Range (Figure 5.14*d*, see also Section 3.5). The subsidence, which can be preceded by a short term uplift of these parts (Figures 5.10*a*, 5.10*b* and Figures 5.11*a* to 5.11*c*; see also Figures 3.7 and 3.9), results in submergence of the Coastal Range and the Longitudinal Valley beneath sea level. The model thus predicts that the Coastal Range, which terminates near Hualien (~24°N), should continue under water further to the north. Both subsidence and down tilting of the northern part of the Coastal Range are corroborated by the geodetic data [*Liu and Yu*, 1990]. The trench itself is not seen in relief. Therefore one can suppose that the trench is filled with sediments rapidly shed by the orogen. In this case, given that the trench is a nonisostatic feature, it should appear distinctly in the free gravity anomalous field as a deep low with amplitude less than -(100-150) mGal [*Watts and Talwani*, 1974]. The gravity map in Figure 5.17 shows such a feature to be present. There is a linear gravity low (dotted in Figure 5.17) off north-eastern Taiwan with an amplitude of ca. -240 mGal. Another east-south stretching deep gravity low of similar amplitude, seen in Figure 5.17, is associated with the Ryukyu subduction zone (see Section 4.9). It is interesting to note that the northern and southern boundaries of the gravity low along north-eastern Taiwan coincide pretty well with the boundaries of the seismicity swarm discussed above (see Figure 5.8).

5.5.6. Toward a More Adequate Geodynamic Model of Taiwan

We proposed above an idealised evolutionary model for an extremely complicated collision in Taiwan. The model is based on both the general geodynamic (mechanical) consideration and simple 2-D modeling. We tried to consistently incorporate into the model all major phases of the evolution (each of which is poorly constrained) and to find a geodynamically consistent explanation for major tectonic features. The reality of course is more complicated and 3-D at lest in the northern part of the collision zone. Our model suggests consumption of the Philippine Sea Plate under Taiwan in a simplistic form of underthrusting (subduction) of this plate similar to what we observe in classical subduction zones. In fact, the whole triangular corner area between the Ryukyu subduction zone and Taiwan seems to be involved in deformation (Figure 5.18). The position of north-western continuation of the Ryukyu zone is unknown. It can only be supposed that the zone continues in a way shown in Figure 5.18 by thick or thin dashed lines. One can see from Figure 5.18 that the proposed southward propagating subduction zone (located off north-eastern Taiwan) is partly or completely covered (hidden) by the frontal part of the overriding Ryukyu arc Plate. This might be the reason why the associated trench is not seen in relief.

We supposed that the Philippine Sea Plate is homogeneous (except the volcanic arc area) and that the collision process is quasi steady state, but this may be not the case. For example, it is not necessary that the conditions remained constant throughout the collision. At the beginning of this process (in the northern part of Taiwan) failure of the lithosphere in the Luzon arc could have occurred immediately along the westward dipping thrust zone without closing the fore arc basin and, hence, without formation of the Longitudinal Valley. The westward subduction of the Philippine Sea Plate could then have ceased after collision at depth of the leading wedge of this plate with the subducted Asian Plate. This could result in intraplate deformation within the mentioned corner zone. Furthermore, the

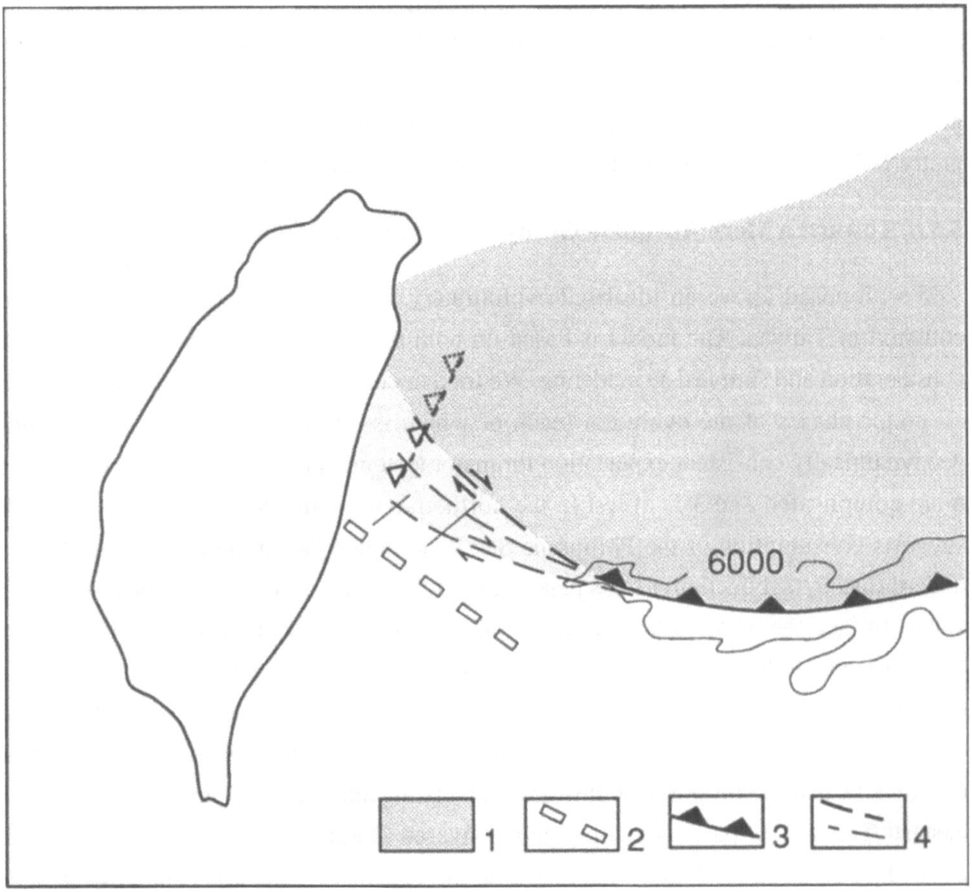

Figure 5.18. Position of main tectonic units to the east of Taiwan
1, overriding Ryukyu arc Plate; 2, the axis of the seismicity zone seen in Figure 5.8 (see text); 3, the Ryukyu subduction zone; 4, possible continuation of the Ryukyu zone (the front of shear-thrust deformation of young sediments overriding the north-westward moving Philippine Sea Plate).

arc at that time (at the beginning of the collision) could be very young [*Barrier and Angelier*, 1986] (or did not exist at all) and hence the lithosphere of the arc be strong. Instead of lithospheric failure and underthrusting of the Philippine Sea Plate, this plate could fail somewhere in the corner zone forming, for example a transform plate boundary. In Figure 5.8 one can recognize a linear zone of concentrated seismicity, abutting Taiwan in its central part from the east. This zone which is schematically shown in Figure 5.18 is oriented nearly parallel to the direction of the plate convergence and is characterised by relatively deep earthquakes (up to 100 km depth). This is definitely a fault zone cutting the lithosphere through the entire thickness. What is the origin of this zone? It may be a result of the different stress state of the Philippine Sea Plate from both sides of the zone. To the north, compressive stress within the plate is released through underthrusting under Taiwan, while to the south the westward motion of this plate is blocked. The fault zone in question can thus accommodate a left-lateral shear between different parts of the Philippine Sea Plate. On the other hand one can suppose this zone to be a right-lateral shear fault assuming that incipient subduction under the north-eastern Taiwan is already blocked (or if it never occurred at all, as discussed above) and the lithosphere to the south of the fault keeps moving westward producing shortening in the Longitudinal Valley - Coastal Range (island arc) area. In any case, judging by the depth of the seismicity, it is necessary to suppose a thrust (subduction) component along this zone. Is this zone a plate boundary? Does this zone connect with another boundaries, with the Ryukyu zone or with the proposed north-east trending incipient subduction zone? Future geological and geophysical investigations will provide insight into these questions. They can also be addressed by 3-D physical modeling.

Another problem is the mechanism for the formation of the metamorphic complex of the Central Range which must be incorporated into the evolutionary model. In our model, the basement of this range is the frontal wedge of the overriding Philippine Sea Plate (see Figure 5.14). For simplicity we have "put" all the metamorphics of the Central Range into this wedge considering that part of the wedge belonged to the Chinese continental margin some time ago and then during complicated evolution of this margin (see [*Lu and Hsu*, 1992, *Teng*, 1992]) was emplaced within the frontal part of the overriding Philippine Sea Plate. In fact, these rocks could have been part of the Asian margin prior to the collision and have been uplifted during the orogeny. The mechanism could be thrusting in the continental crust of the subducted Asian Plate similar to that modeled in the next chapter. To test this mechanism experimentally it is necessary to allow the subducted continental crust of the Asian Plate to be scraped from the mantle layer by making the lower crust weaker (see Chapter 6). In the experiments described above, the continental crust is strongly coupled with mantle layer and cannot be scraped from it.

5.6. Conclusions

A young age of the subducting plate, a high rate of subduction, the subduction of different asperities, and collision with plateaux, microcontinents, and continents result in compression of the lithosphere in the subduction zone. Compressive stresses can locally exceed the lithospheric yield limit within both subducting and overriding plates, depending mainly on where the lithosphere is the weakest. The yield limit can be reached in the back arc area, directly in the volcanic arc, behind the outer rise (within the subducting plate) or far from the subduction zone if compression is caused by a large scale collision. Buckling, localization of deformation and failure of the lithosphere develop at that place. Failure occurs along one of two thrust faults dipping in opposite directions. New subduction zone with appropriate polarity forms at the fault, with subduction in the old zone ceasing. If the plate fails in the arc area then the resultant fault dips under the arc from one of its sides. When the fault dips in the direction opposite to the Benioff zone then plate failure results in a jump of the subduction zone and a subduction reversal. Initial stage of such a process is probably going on currently in the Kuril and Japan zones. If failure occurs along a fault dipping under the arc in the same direction as the Benioff zone then underthrusting of the frontal segment of the overriding plate and its consumption in the mantle follow the failure of this plate. The trench migrates toward the arc while the arc retreats during this process and then subduction can stabilize again. This is a kind of rapid erosive pulse in the overriding plate which does not change the general tectonic plan of the plate boundary. It is not clear whether such processes occurs in nature or not. More probable is the underthrusting of the fore arc area beneath the arc only to some depth after which failure occurs along another fault dipping under the arc from the opposite side. Such a process seems to occur in the Luzon arc in the Taiwan area where oblique collision of the arc with the Asian continental margin has developed in the last few million years. In the initial stages of the collision, the most intensive deformation and thrusting develop along the zone dipping under the arc to the east. Underthrusting along this zone (which is the proto-Longitudinal Valley Fault) leads to a closing of the fore arc basin. The western edge of the overriding volcanic block rises up above sea level forming the Coastal Range with the suture zone in front of it, which is the Longitudinal Valley. This process is unstable and ceases after some time. Increasing compressive stress causes concentration of the deformation along the westward dipping thrust zone. Failure of the lithosphere along this zone initiates westward (west-northward) underthrusting of the Philippine Sea Plate under the north-eastern margin of Taiwan. Due to the oblique collision, this process propagates southward at a rate near 10 cm/yr. Further subduction of the Philippine Sea Plate causes a collision at depth of both subducted Asian and Philippine Sea Plates, leading to an increase in compression. This might result either in reactivation of the overthrusting of the Taiwan orogen on the Asian Plate or further reorganization of the plate boundaries to the east of Taiwan.

CHAPTER 6

CONTINENTAL SUBDUCTION

6.1. Introduction

The main difference between mechanics of the classical subduction of oceanic lithosphere and the subduction (consumption) of continental lithosphere in the collision zones beneath either continental or oceanic lithosphere, is defined by the presence of a thick continental crust in the latter case. High buoyancy of the crust prevents it from being pushed or dragged down deep into the mantle. The general picture of the continental lithosphere subduction (continental subduction [*Mattauer*, 1986]) seems to depend first of all on two major factors; the effective strength of the crust and the coupling between the crust and the mantle layer of the lithosphere. If both the strength and the coupling are high enough, one can expect the style of subduction to be similar to that shown in Figure 4.3 or Figure 4.4 (experiments on subduction of a low density lithosphere). If the effective strength of the crust is too low, then independently of the coupling between the crust and mantle, the crust would not be dragged down to a large depth, but be scraped off the mantle layer of the plate. The result would be the formation of a huge crustal accretionary prism, or orogen, similar to but much larger than sedimentary prisms in classical oceanic subduction zones. A more complicated and probably realistic situation corresponds to the case of comparatively strong crust (at least the upper crust) and a low coupling with the mantle. The low coupling may be due to the presence of a weak ductile layer, or "crustal asthenosphere", in the lower crust [*Ranally and Murphy*, 1987]. In this chapter we present results of a simple modeling of continental subduction with the lithosphere comprising "crustal asthenosphere".

6.2 Modeling Approach

Two-dimensional modeling is performed with a three-layer lithospheric model which includes two crustal layers (the lower layer is very weak) and a mantle part [*Shemenda and Grocholsky*, 1992]. An effectively fluid low-viscosity asthenosphere underlies the lithosphere. The scheme of the experiments is depicted in Figure 6.1 (we use the same experimental installation as for the previous 2-D experiments).

Figure 6.2*a* shows the maximum differential stresses plotted against depth for a typical continental lithosphere. These stresses can be treated as effective lithospheric yield limits at appropriate depths under strain rates that are typical of geological processes. Taking these data into account, the yield limit for the model lithosphere can be approximated as shown in Figure 6.2*b*. We assume that the upper crustal layer and mantle layer of the lithosphere possess elastico-semi brittle and elastico-semi plastic properties, respectively. Both layers

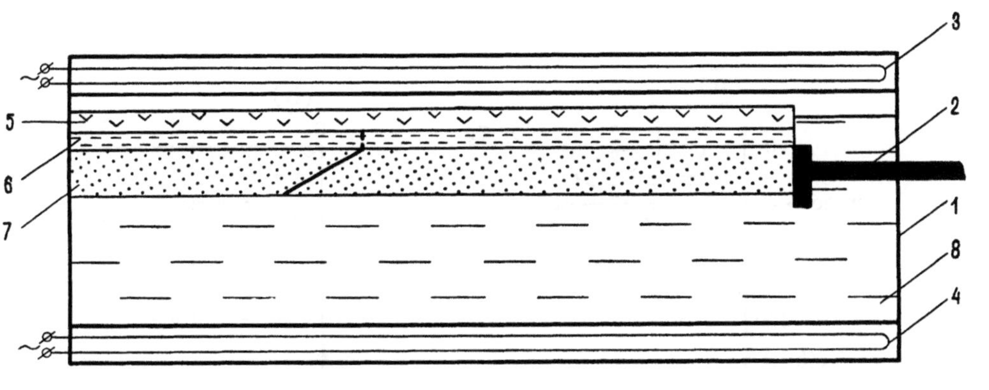

Figure 6.1. Scheme of the experiment and the experimental installation
1, tank; 2, piston; 3 and 4, heaters; 5, upper (first) crustal layer; 6, lower (second) crustal layer; 7, mantle part of the lithosphere (third layer), 8, asthenosphere.

have approximately the same effective yield limit which probably should be near that obtained for the oceanic lithosphere in Chapter 3. The lower crust, the ductile "crustal asthenosphere" may have yield limit 1 to 3 orders less than that for the upper crust and the underlying mantle.

The subduction of the oceanic lithosphere which precedes continental subduction (collision) is not modeled. The process starts directly with the collision of continental plates. The collision asymmetry inherited from the oceanic subduction is incorporated by making an inclined cut in the mantle layer of the lithosphere (see Figure 6.1). The cut defines a subducting and an overriding plates.

As in the previous experiments the driving mechanism for the subduction (collision) is horizontal compression of the lithosphere by the piston. We investigate two cases. In the first, the piston acts only on the mantle layer of the plate as shown in Figure 6.1 allowing sliding of the upper crustal layer over the crustal asthenosphere. In the second case the piston pushes all the lithospheric layers, so that relative sliding of the layers is absent, or only local.

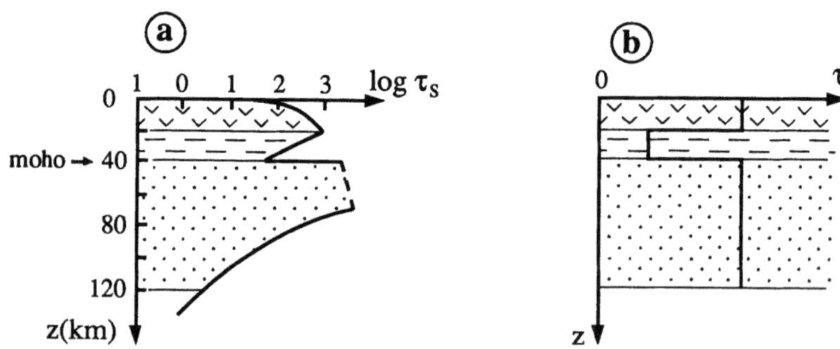

Figure 6.2. Effective yield limit τ_s in the lithosphere (the differential stresses at a constant strain rate) as a function of depth
(*a*) Actual lithosphere, (*b*) Model lithosphere.

6.3. Similarity Criteria and the Model Materials

The similarity criteria for modeling the continental subduction in the framework of the above scheme can be written according to (1.43) and (1.44) as

$$\begin{gathered}\tau_{sl}/(\rho_l g H_m) = \text{const}, \quad \tau_{s1}/(\rho_{c1} g H_m) = \text{const}, \quad \tau_{s2}/(\rho_{c2} g H_2) = \text{const}, \\ E_l/\tau_{sl} = \text{const}, \quad E_l/E_{c1} = \text{const}, \quad \rho_l/\rho_a = \text{const}, \\ \rho_l/\rho_{c1} = \text{const}, \quad \rho_{c1}/\rho_{c2} = \text{const}, \quad H_m/H_{c2} = \text{const}, \\ H_{c1}/H_{c2} = \text{const}.\end{gathered} \tag{6.1}$$

Here τ_{sl}, τ_{s1},τ_{s2} are the shear yield limits for the mantle lithospheric part, for the upper and lower crustal layers, respectively; E_l and E_{c1} are the modules of elasticity for the mantle and for the upper crustal layers, respectively; ρ_l, ρ_a, ρ_{c1}, ρ_{c2} are the densities of the mantle lithospheric layer, asthenosphere, upper and lower crustal layers, respectively; H_m, H_{c1}, H_{c2} are thicknesses of the mantle lithospheric layer, upper and lower crustal layers; g is the acceleration of gravity, and superscripts "o" and "m" denote as before original (nature) and model parameters, respectively.

We assume the following parameter values for the prototype which are within the available estimates: $\tau^o_{sl} \approx \tau^o_{s1} = 1.4 \times 10^8$ Pa; $\tau^o_{s2} \approx 4 \times 10^6$ Pa; $E^o_l = E^o_{c1} = 1.4 \times 10^{10}$ Pa; $\rho^o_l \approx \rho^o_a = 3.3 \times 10^3$ kg/m^3; $\rho^o_{c1} \approx \rho^o_{c2} = (2.8\text{-}2.9) \times 10^3$ kg/m^3; $H^o_m = 5 \times 10^4$ m; $H^o_{c1} = H^o_{c2} = 2 \times 10^4$ m. Taking into account these values, criteria (6.1) can be satisfied using the following model materials at the working temperature 40°C (see Section 2.3):

Compound N1 ($\tau^m_{sl} = 11$ Pa, $E^m_l \approx 1.1 \times 10^3$ Pa and $\rho^m_l = 10^3$ kg/m^3) for modeling the mantle layer of the continental lithosphere.

Compound N2 ($\tau^m_{s1} = 10$ Pa, $E^m_{c1} \approx 10^3$ Pa and $\rho^m_{c1} = 0.86 \times 10^3$ kg/m^3) for modeling the upper crustal layer.

Compound N3 ($\tau^m_{s2} = 0.41$ Pa , $\rho^m_{c2} = 0.86 \times 10^3$ kg/m^3) and

Compound N4 ($\tau^m_{s2} = 0.28$ Pa, $\rho^m_{c2} = 0.86 \times 10^3$ kg/m^3) for modeling the crustal asthenosphere.

As in the previous experiments, pure water of density $\rho^m_a = 10^3$ kg/m^3 is used to model the asthenosphere.

Above values for the model parameters satisfy the similarity criteria (6.1) if the layers of the lithosphere in the model have the thicknesses $H_{\mathrm{m}}^{\mathrm{m}} = 1.3 \times 10^{-2}$ m, and $H_{\mathrm{c}_1}^{\mathrm{m}} = H_{\mathrm{c}_2}^{\mathrm{m}} = 0.5 \times 10^{-2}$ m.

6.4. Experimental Results

Several dozen experiments have been carried out. Below we describe the most typical ones.

Experiment 6.1. The lower crustal layer in this experiment is modeled by compound N3 and is partially (weakly) coupled with the upper layer. The lower (mantle) layer is pushed to the left by the piston as shown in Figure 6.1. The result is presented in Figure 6.3. At the initial stage of the experiment (Figure 6.3*b*) the mantle part of the subducting plate is subducted under the overriding plate. The lower crustal layer (the central layer in Figure 6.3) is dragged down and partially scraped off. The crust is progressively thickening due to a "pumping" of the "crustal asthenosphere" into the collision zone, with the front of the thickening migrating to the right (Figure 6.3*c* to 6.3*f*). The upper crustal layer experiences only vertical movements which are a passive response to the pressure from below.

Thickening of the crust results in increasing sagging of the mantle layer of the subducting plate and also, in the growth of topography above the thickened zone. Progressive pulling of the second layer under the overlapping wedge of the overriding mantle, causes gradual flattening (beginning in Figure 6.3*d*) of the mantle layer of the subducted plate until it reaches the bottom of the overriding plate (Figures 6.3*e* and 6.3*f*).

The subvertical white dashed line in the second crustal layer in Figure 6.3*f*, shows the boundary between the lower crustal layers initially belonging to different plates (in fact, the layers have different colours in the two plates). The position of this boundary remained practically unchanged compared to its initial position before the experiment (see corresponding dashed line in Figure 6.1). This means that there was almost no intrusion of the crustal asthenosphere of the subducting plate into that of the overriding plate during subduction.

Experiment 6.2 differs from the previous experiment in that the second layer of the overriding plate is made of a weaker material, compound N4. Due to this modification some intrusion of the second crustal layer of the subducting plate into the overriding plate has occurred (Figure 6.4), the frontal part of the mantle layer of the overriding plate underwent down bending, and less amount of the second crustal layer material has been dragged into the interplate contact zone. No flattening of the subducted plate segment

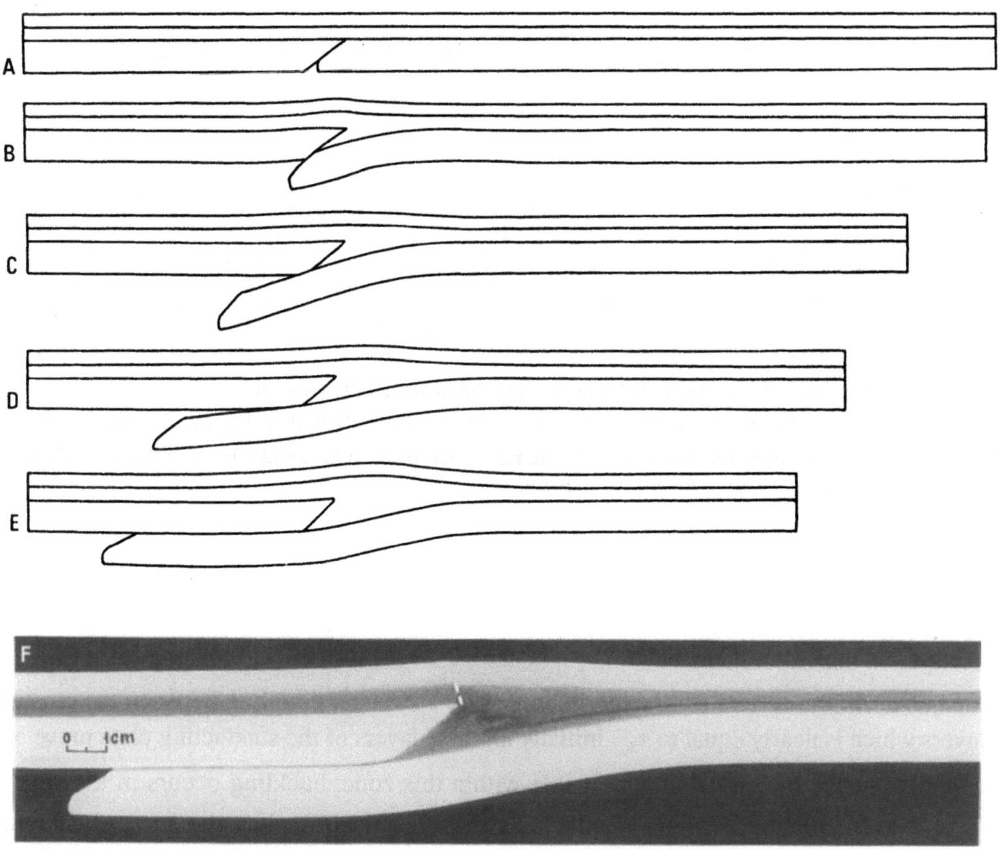

Figure 6.3. The result of experiment 6.1 ($\tau^m_{sl} \approx \tau^m_{s1} \approx 11$ Pa; $\tau^m_{s2} = 0.41$ Pa; $E^m_l \approx E^m_{c1} \approx 1.1 \times 10^3$ Pa; $\rho^m_l = \rho^m_a = 10^3$ kg/m^3; $\rho^m_{c1} = \rho^m_{c2} = 0.86 \times 10^3$ kg/m^3; $H^m_m = 1.3 \times 10^{-2}$ m; $H^m_{c1} = H^m_{c2} = 0.5 \times 10^{-2}$ m; the coupling between the crustal layers is weak; the piston acts only on the plate mantle layer) (*a*) to (*e*) The stages of model process (drawn copies of photos of the model). (*f*) The photo of the model section which is made after freezing the deformed "lithosphere" (the black deformed line in the central part of the "crustal asthenosphere" of the subducting plate is the thin dyed layer which was formerly (before deformation) horizontal and parallel to the plate bottom and surface; the white dashed line in the second layer in the collision zone shows the boundary between the second layers of both plates (before experiment the line was vertical, see Figure 6.1)).

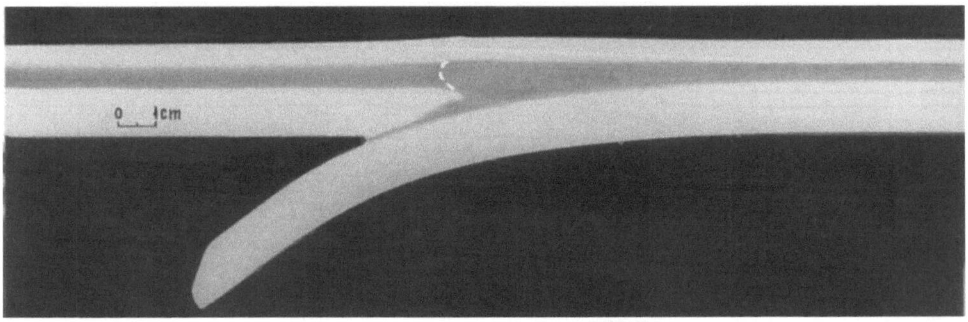

Figure 6.4. The result of experiment 6.2 (the photo of the frozen model section).
For the subducting plate τ^{m}_{s2}= 0.41 Pa, for the overriding plate τ^{m}_{s2}= 0.28 Pa; the coupling of the crustal layers in the subducting plate is weak, the piston acts only on the mantle layer of the lithosphere (see for additional explanations the legend to Figure 6.3).

occurred in this experiment.

Experiment 6.3 differs from experiment 6.1 by a stronger coupling between the crustal layers which is nearly equal to τ_{s2}. Initially all the 3 layers of the subducting plate move as a whole outside the collision zone while within this zone, buckling occurs in the upper crustal layer (Figure 6.5). Then mantle layer of the subducting plate also looses buckling stability and fails near the piston along an inclined fault (Figure 6.5*b*). Simultaneously, a shear flow begins in the crustal asthenosphere (τ_{s2} is exceeded) resulting in decreasing the lithospheric compression. The continuation of deformation directly in the collision zone is similar to experiment 6.1. Thus, due to a strong coupling of the layers in this experiment, compressive stresses initially reached the yield limit in both the upper crust and the mantle part of the subducting plate but the failure of the material occurred then only in the mantle layer resulting in origination of a new subduction (collision) zone. This result demonstrates the inadequacy of the boundary conditions used in the model. Instead of having failure of

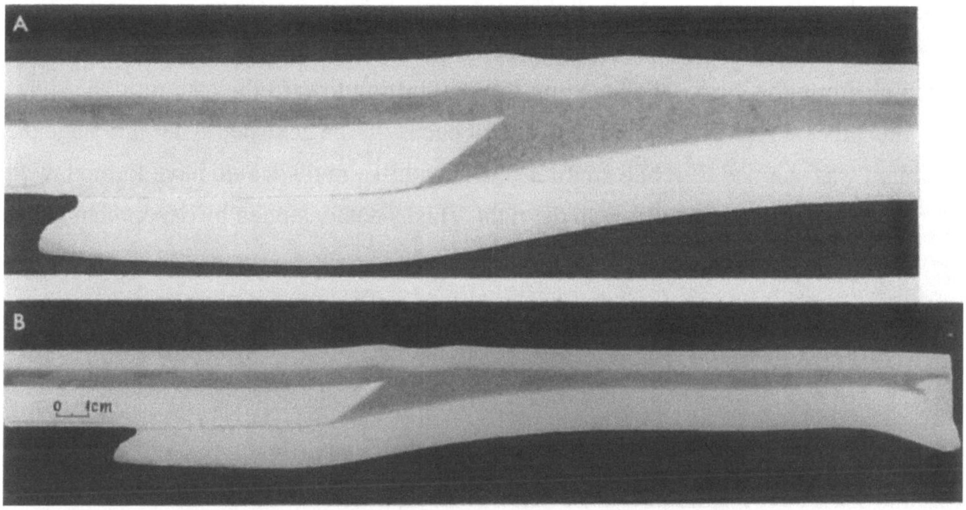

Figure 6.5. The result of experiment 6.3 (section of the frozen deformed model)
τ^{m}_{s2} = 0.41 Pa for both plates, the coupling of the crustal layers is strong, the piston acts only on the mantle layer of the plate.
(*a*) The collision zone. (*b*) A general view of the model (see also the legend to Figure 6.3).

the upper crustal layer which seems to be more realistic we obtained failure in the mantle layer. The reason being the concentration of stresses in the plate near the piston due to its rigidity. In reality, the lithosphere experiences a "soft" pushing load due to either/both tangential tractions distributed over the lithospheric base or/and sliding of the lithosphere from mid ocean ridges (ridge-push). Under such conditions a concentration of stresses, obtained in the model, does not occur. To overcome this problem and more closely approximate the actual situation (i.e. to have failure of the upper crust occur before stresses reach the yield limit in the mantle layer of the lithosphere) one can either make the upper crustal layer weaker or apply pressure from the piston to the whole lithosphere, not only to its mantle layer. Both variants have been tested. The results are similar. Those described below were obtained using the latter method in which the piston acts on the entire plate section.

Experiments 6.4 and 6.5. The conditions are the same as in experiment 6.3, except for the position of the piston, it now pushes the plate across the entire section. As expected, failure and thrusting in the upper crustal layer follow its buckling (Figures 6.6 and 6.7). A first crustal thrust fault occurs in the deepest flexural sag, with local "crustal subduction" developing along this fault. The buckling front is shifting toward the piston (to the right) during this process and then another thrust fault occurs in the next sag. After this stage the experiment was stopped. If it had continued, a new thrust faults would have formed, with front of thrusting gradually shifting to the right. This is demonstrated by experiment 6.5 in Figure 6.8.

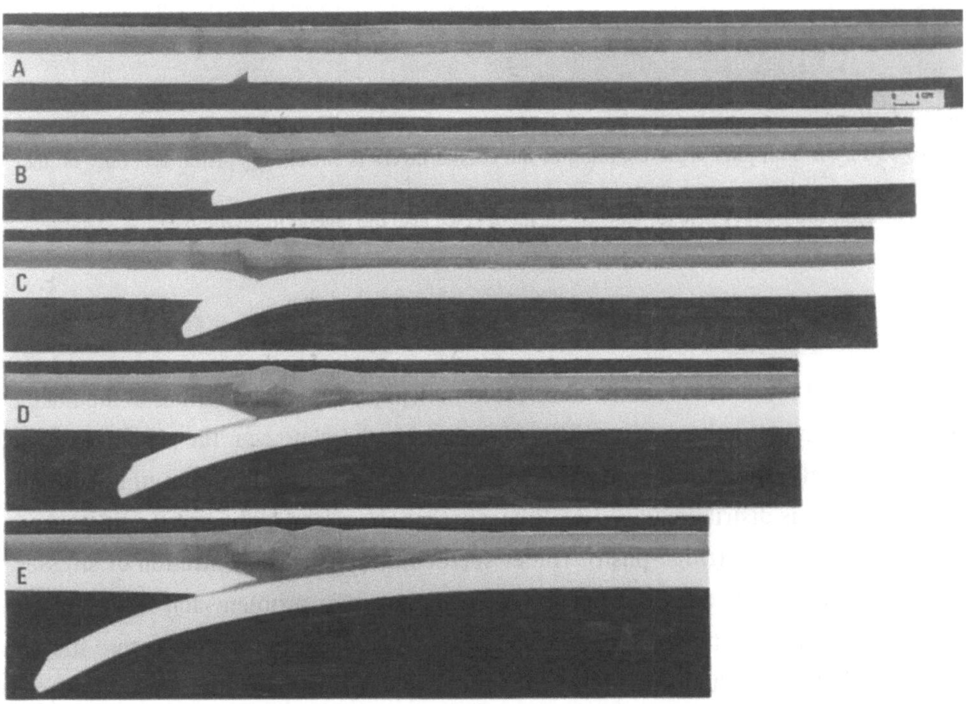

Figure 6.6 (see continuation on the next page)

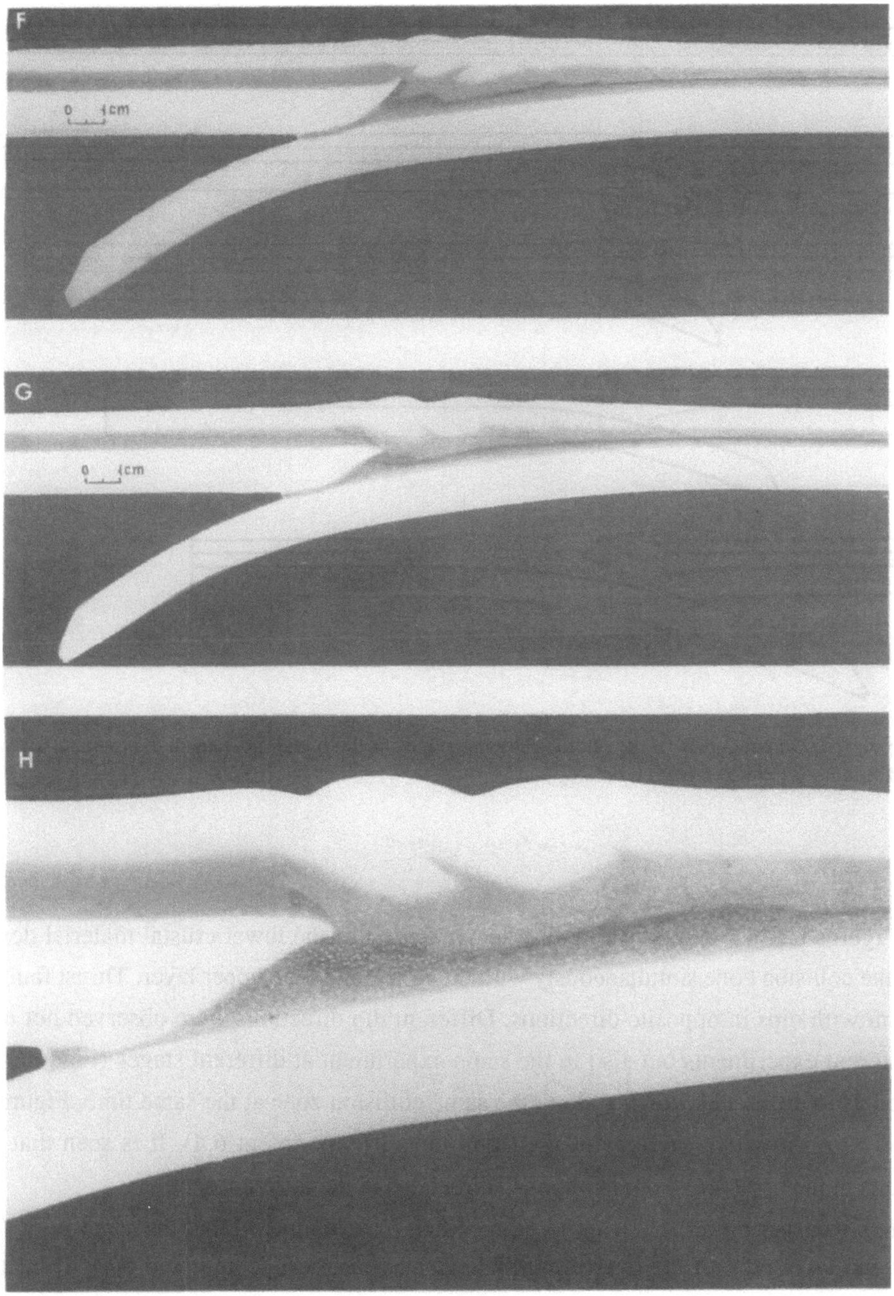

Figure 6.6. The result of experiment 6.4. The parameter values are the same as in experiment 6.1 (Figure 6.3), the piston pushes the whole plate section.
(*a*) to (*e*) Photos showing stages of the model process. (*f*) and (*g*) Sections of the frozen deformed model. (*h*) Enlarged view of the collision zone corresponding to the section in Figure 6.6*g* (see legend to Figure 6.3).

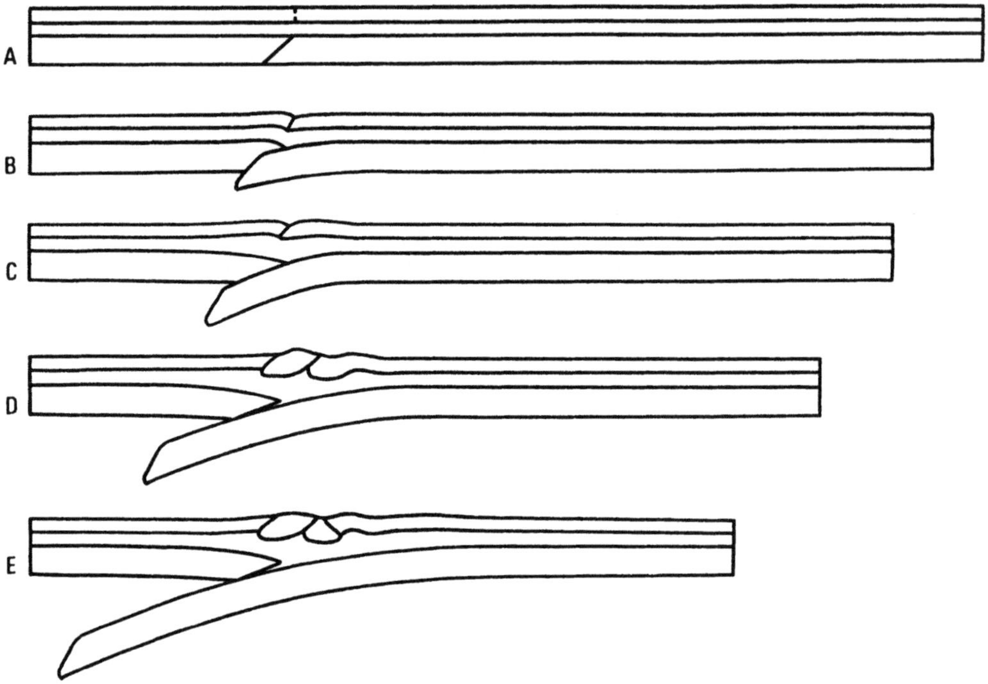

Figure 6.7. Drawn copies of the photos showing stages of the model process in experiment 6.4 which correspond to Figures 6.6*a* to 6.6*e*.

Figures 6.6, 6.7 and 6.8 show that accumulation of the lower crustal material develops in the collision zone simultaneously with the thrusting in the upper layer. Thrust faults may form with dips in opposite directions. Different dip directions were observed not only in different experiments but also in the same experiment at different stages (Figure 6.8), as well as in different segments along the same collision zone at the same time. Figures 6.6*f* and 6.6*g* show two sections of the same model (experiment 6.4). It is seen that thrust faults in the upper crust dip in opposite directions in the different sections.

Experiments 6.6 differs from two preceding experiments in that the yield limit for the second layer ($\tau_{s_2}^{m}$) of the overriding plate is 1.5 times less (compound N4). This enables the upper layer of the subducting plate (together with the underlying material of the second layer) to intrude at the initial stage of the experiment some distance into the second layer of

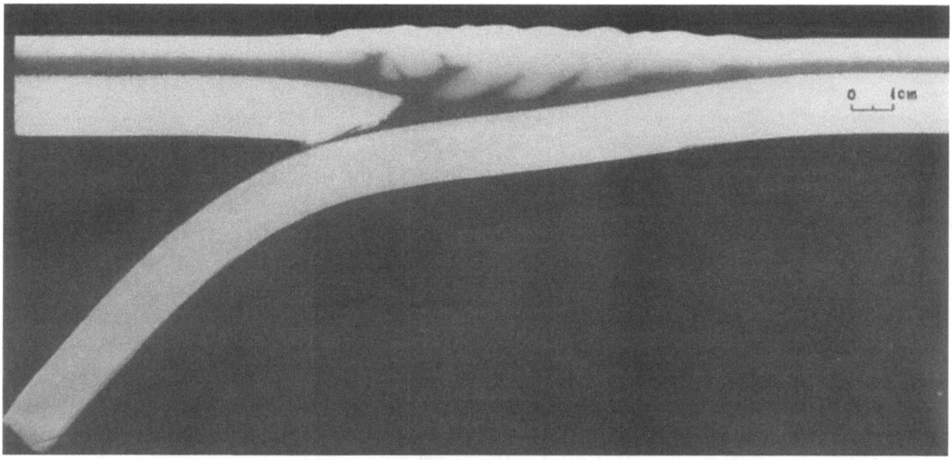

Figure 6.8. Experiment 6.5 (the frozen deformed model section). The conditions are the same as in experiment 6.4 (Figure 6.6), however this experiment was stopped at a later stage.

the overriding plate (Figure 6.9). One can notice that in the initial stages of experiment 6.6 (Figure 6.9) thrust faults dip toward the overriding plate, whereas at the final stages they dip in the opposite direction.

Experiment 6.7. In this experiment the "crustal asthenosphere" is made as thin as 1 mm. The thickness of the upper crust is accordingly increased so that the total crustal thickness remains the same. The style of model deformation in experiment 6.7 (Figure 6.10) does not significantly differ from similar experiments described above. At the same time, unlike experiments 6.4 to 6.6 (Figures 6.6 to 6.9) where the crustal asthenosphere was thicker, in this case a flattening of the subducted mantle layer is observed (Figure 6.10).

Experiment 6.8. In this experiment the crustal asthenosphere is entirely absent in the subducting plate (the coupling between the thick, strong crustal layer and the underlying mantle is at a maximum). The result (Figure 6.11) considerably differs from the previous experiments. In experiment 6.8, full stripping of the crust occurs only in the leading edge of the subducting plate. Further some stripping develops only in the upper crustal part,

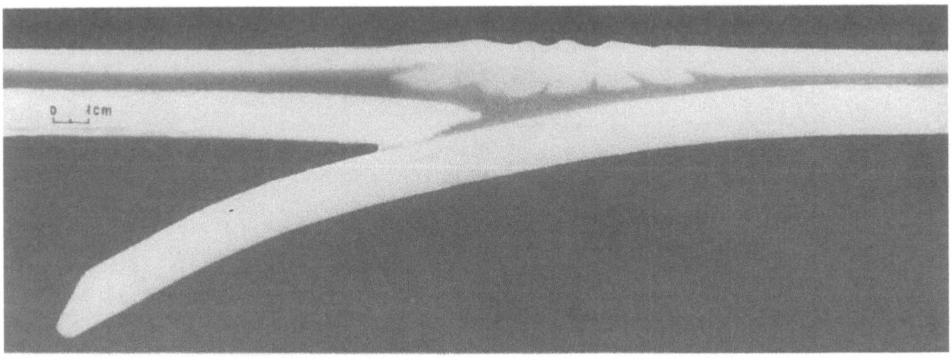

Figure 6.9. Experiment 6.6 (the section of the frozen deformed model). This experiment differs from the previous one in the composition of the second layer of the overriding plate, it is made of the weakest compound N4 ($\tau^{m}_{s_2}$= 0.28 Pa), see the legend to Figure 6.3.

leading to a comparatively small thickening of the crustal layer. The rest of the light crust along with the mantle lithospheric layer is horizontally subducted into the mantle due to high buoyancy of the subducted plate.

Experiments 6.9 and 6.10 (Figure 6.12). The upper crustal layer and the mantle part of the lithosphere in these experiments are made of compounds N2 and N1, respectively. The second layer in both plates is made of the weakest compound N4. The piston acts on the whole plate section. At the initial stages of these experiments, failure of the upper layer occurs in the zone of plates contact, similarly to the previous experiments. The process proceeds depending on the thrust fault dip direction. If the fault dips toward the overriding plate, then the upper layer of the subducting plate intrudes into the second layer of the overriding plate (Figure 6.12*a*). If the dip is opposite, then the upper layer of the subducting plate overrides the surface of the overriding plate (Figure 6.12*b*). The distance of both the intrusion (Figure 6.12*a*) and overthrusting (Figure 6.12*b*) is considerable in these experiments.

In experiments 6.9 and 6.10, similarly to experiments 6.2 and experiments 6.4 to 6.6 (Figures 6.4 and 6.6 to 6.9), the flattening of the subducted plate does not occur, the frontal part of the mantle layer of the overriding plate undergoes subsidence and only a

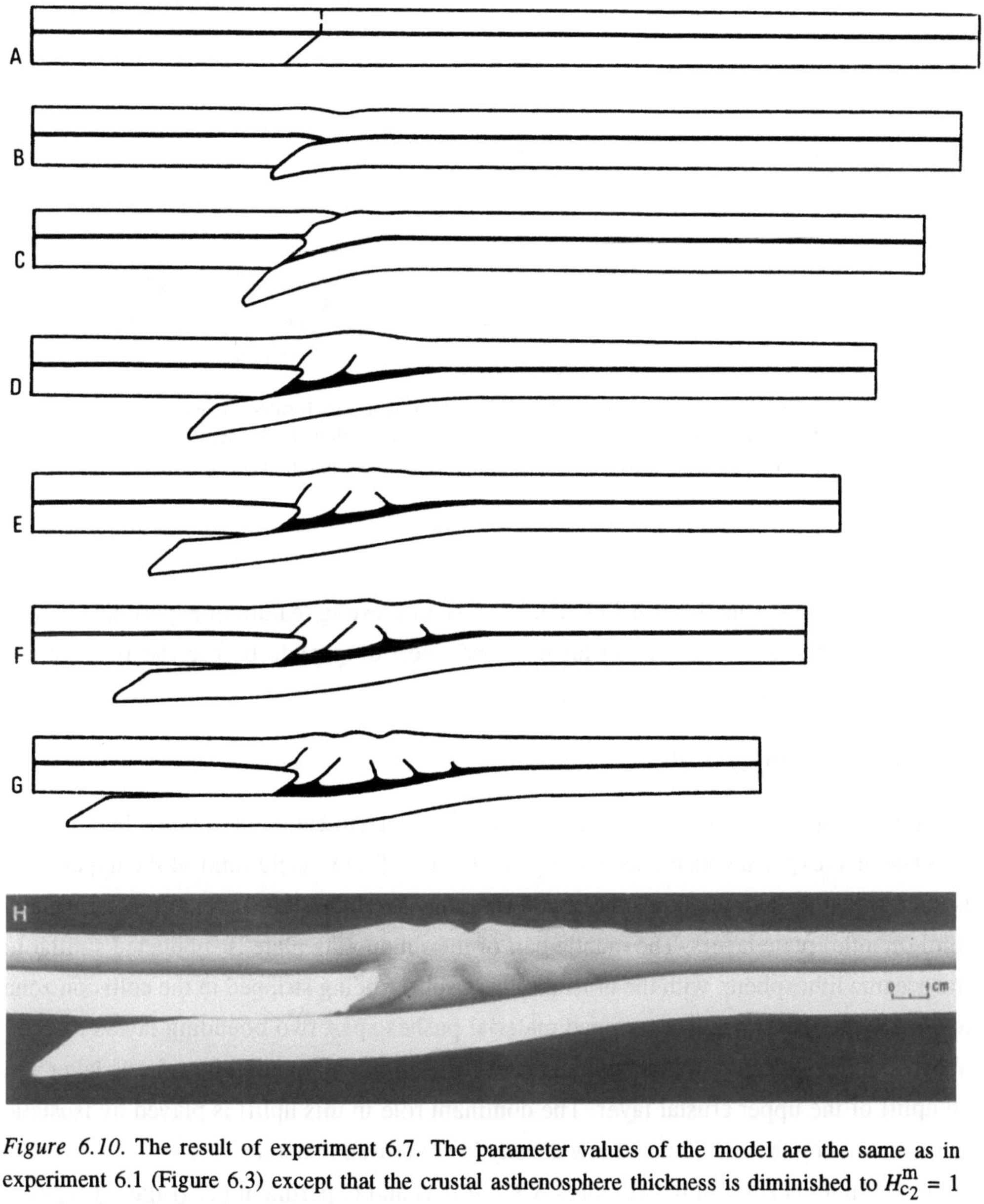

Figure 6.10. The result of experiment 6.7. The parameter values of the model are the same as in experiment 6.1 (Figure 6.3) except that the crustal asthenosphere thickness is diminished to $H_{c2}^{m} = 1$ mm; the piston pushes the whole plate section.
(*a*) *to* (*g*) Drawn copies of the photos (the thin black central layer is the "crustal asthenosphere"). (*h*) The section of the frozen deformed model corresponding to Figure 6.10*g*.

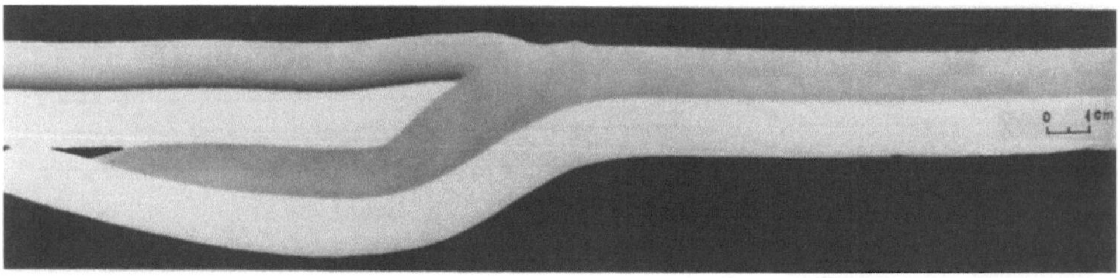

Figure 6.11. The result of experiment 6.8 (the section of the frozen deformed model). The experiment differs from the previous in that the second layer is absent in the subducting plate, the thickness of the first crustal layer in this plate is $H_{c1}^{m} = 10^{-2}$ m, its coupling with the mantle layer is strong.

small part of the second layer is pulled down. An interesting feature of experiment 6.10 is a local thickening of the second layer of the overriding plate before the front of the overriding upper layer (Figure 6.12*b*).

6.5. Analysis of the Results

Let us consider the experimental results from the mechanical point of view. In the first 3 experiments (experiments 6.1 to 6.3, Figures 6.3 to 6.5), the yield limit of the upper crust is not reached, and all the lithospheric deformation develops at the level of the second and third (mantle) plate layers. The mantle part of the subducting plate is subducted similar to the oceanic lithosphere, with the crustal asthenosphere being stripped in the collision zone and partly pulled down. The stripped material pushes apart two bounding layers directly above and below it, producing both a greater sagging of the mantle lithospheric layer and an uplift of the upper crustal layer. The dominant role in this uplift is played by isostatic forces caused by the accumulated low density lower crustal material.

In experiments 6.1 and 6.3 (Figures 6.3 and 6.5) and experiment 6.7 (Figure 6.10) the subducted mantle layer flattens at a particular stage of the collision. This occurs in spite of the fact that the layer density is equal to the density of the asthenosphere. In the other experiments which were carried out under slightly different conditions, but with the same

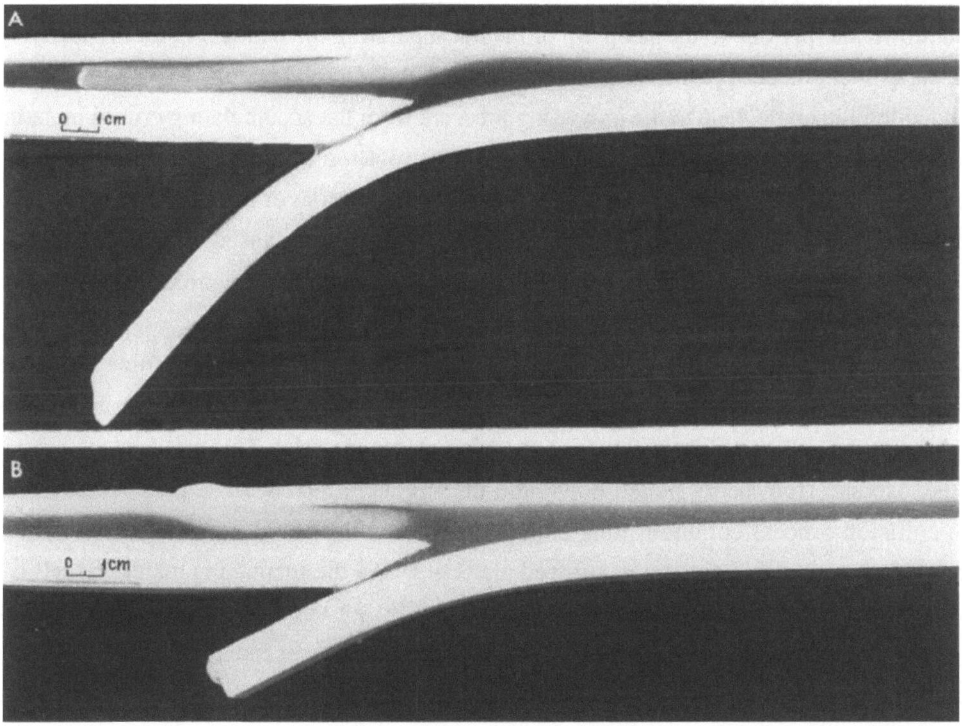

Figure 6.12. The results of experiments 6.9 (*a*) and 6.10 (*b*) (the sections of the frozen deformed models). The conditions of these experiments are the same; the second layer in both plates made of the weakest compound N4 ($\tau^{m}_{s_2}$ = 0.28 Pa), the piston pushes the whole plate section.

mantle layer of the lithosphere (experiments 6.2, 6.4 to 6.6, 6.9 and 6.10), such a plate flattening was not observed. The effect of flattening can be explained as follows. The stripped material of the second layer in experiments 6.1, 6.3, and 6.7 does not penetrate into the region above the mantle layer of the overriding plate and hence does not produce additional pressure on the surface of the overriding mantle layer. The frontal wedge (its mantle part) of the overriding plate appears to be floating in the stripped material and is slightly upturned under its pressure. Initially the mantle layer of the subducting plate was bent down sharply due to the steep dip of the cut separating the plates (Figure 6.1). During the experiment the stripped crustal material widens the interplate boundary and lowers the subduction dip angle. Due to the increasing pressure from the accumulating crustal material the subducted mantle layer experiences clockwise rotation (in cross section) around the deepest point of the interplate boundary up to reach the bottom of the overriding plate. An additional factor which acts in the same direction is the torque caused on the one hand by the friction at the surface of the subducting mantle layer in collision zone, and on the other hand by the force pushing this layer.

In the experiments where the crust of the subducting lithosphere is intruded into the crustal asthenosphere of the overriding plate, the mantle part of this plate undergoes additional sagging under the pressure of the overlapping material. This prevents the crustal asthenosphere from being pulled down and the subduction from flattening. However, if the collision process continues long enough, the flattening effect would still occur after accumulation of a large amount of stripped material above the subducting mantle layer.

Experiments 6.1 and 6.2 demonstrate the mechanism for the differential deformation of the crustal layers as well as the possibility of an increase in the crustal thickness in collision zones due to accretion of only (or mainly) the lower crust. Such a scenario requires the upper crustal layer of the whole subducting plate to slide for a great distance over the mantle lithospheric layer.

The experiments in which relative sliding of the lithospheric layers has been inhibited out of the collision zone seem to be more realistic. In experiments 6.9 and 6.10 (Figure 6.12) the crustal asthenosphere in both plates is modeled by the material of the lowest strength. Deep penetration (overlapping) of the upper crustal layer into (over) the overriding plate occurs in these experiments due to low resistance from the second layer (in experiments 6.9 and 6.10 the second layer is really like a low-viscosity asthenosphere). In experiments 6.4 to 6.7 (Figures 6.6 to 6.10) the crustal asthenosphere (either in both plates, or only in the subducting plate) consists of stronger material. Such a material provides a higher resistance for the intrusion of the upper crust into the crustal asthenosphere and thereby a quasi cyclic character of the continental subduction. The compressive stresses in the lithosphere reach the maximum just before failure of the upper crust (formation of a major crustal thrust fault) and then drop during the failure. The failure is followed by localized deformation (underthrusting) along the fault and crustal subduction, with deformation outside of the fault zone ceasing. During crustal subduction compression increases once again and deformation involves ever wider area of the crust and the lithosphere as a whole. After the formation of a next major crustal thrust fault the cycle is repeated.

The subducting lithosphere moves as a whole outside the collision zone while within this zone the different layers deform in different ways. The upper layer is broken by a set of thrust faults and the lower crustal layer experiences plastic shortening and accretion, the same as in experiments 6.1 to 6.3 (Figures 6.3 to 6.5). Thickening of the crust in the collision zone due to both these processes occurs proportionally to layer thickness (the thicker the layer, the greater its contribution to the crustal thickening).

The crustal asthenosphere thickness does not significantly affect the style of crustal deformation. The thickness may be even very small as in experiment 6.7 (Figure 6.10). It is only important that the coupling between the strong crustal layer and the mantle layer of the lithosphere be weak enough. If the lubrication between these layers is absent (as in experiment 6.8, Figure 6.11), the style of deformation is changed because the crust is dragged to a great depth.

6.6. Concluding Remarks and Possible Application of the Results to the India-Eurasia Collision

Which of the modeled situations is more realistic? None of the experimental settings seem to be entirely satisfactory. In experiments 6.1 and 6.2 the upper crustal layer of the whole subducting plate slides for a great distance over the mantle lithospheric layer which is hardly realistic. In experiments 6.9 and 6.10 (Figure 6.12) one large-amplitude overthrust fault forms causing considerable crustal thickening within the overriding plate far from the convergent boundary. Such a mechanism appears to be rather tempting for explaining the crustal thickening in the hinterland of the India-Eurasia collision zone in Tibet. However, formation of a single major crustal thrust in the collision zone is hardly consistent with geological evidence.

The results of experiments 6.4 to 6.7 seem to be more reasonable, especially the successive (propagating) thrusting in the upper crust forming the upper part of the orogen.

These experiments probably represent a good approximation to continental collision initiated within the continental lithosphere. However it is not obvious whether these results can be applied directly to the continental subduction which follows the subduction of the oceanic lithosphere. An inclined cut in the mantle layer of the lithosphere made before the experiments (see Figure 6.1) might not be sufficient to properly imitate an inheritance of the collision from the subduction of the oceanic lithosphere. The continental crust in the transition zone from the ocean to the continent is thin and probably does not contain crustal asthenosphere. Therefore at least frontal segments of the continental crust must be subducted during collision into the mantle (without being scraped in the interplate zone) following the oceanic lithosphere to some depth. Sooner or later the continental crust will be scraped due to both the growing buoyancy force and the heating of the subducted crust. However the delay in time of this scraping may have important geodynamic and geologic consequences. The delay will influence, for example the position (dip angle) of the Benioff zone (subducted mantle lithospheric layer), the mechanisms for the orogen uplift and exhumation of the rocks metamorphosed at great depth. Future modeling of continental subduction must thus incorporate the stage of the oceanic subduction preceding the collision.

Interesting ideas about dynamics of the collision follow from experiment 6.3 (Figure 6.5). The buckling in this experiment occurs simultaneously in both the upper and lower plate layers. In the upper layer the buckling is located directly in the collision zone and does not propagate outside due to the resistance (friction) from the underlying crustal asthenosphere which is thin and sufficiently strong. Scaling up the experimental results to match the prototype, yields a buckling wavelength length λ_c of about 60 km. In contrast to this, bending of the mantle layer propagates over the entire plate length due to a low viscosity (strength) of the mantle asthenosphere (water). The wavelength of this deformation corresponds to λ_l = 300-400 km in nature. It is interesting that bending of the mantle layer is practically completely suppressed in the crustal asthenosphere and is not transmitted to the surface. If the colliding plates contain a thinner oceanic lithosphere, then buckling would most probably develop in that area. Due to the absence of a crustal asthenosphere in the oceanic lithosphere, its bending would be displayed in relief. As was discussed above, the upper crust in experiment 6.3 did not fail due to a stress concentration and the failure of the subducting plate near the rigid piston which does not correspond to nature. In the case of more realistic "soft" loading of the lithosphere (due to mantle convection or/and sliding of the lithosphere away from mid ocean rises) the buckling of both the upper crust and the mantle layer (or the mentioned oceanic segment belonging to one of the interacting plates) can be followed by failure of the crust in the collision zone. The failure will cause active thrusting (crustal subduction) along the fault, a drop in lithospheric compression, and thus a cessation of "intraplate" deformation (buckling and thrusting) within the oceanic (thin) segment of the interacting plates. During crustal subduction (which is one of the elements of orogeny), compression increases once again, resulting in both a cessation of this subduction (orogeny) itself and in a reactivation of the intraplate deformation in the oceanic lithosphere. Such a cycle can repeat several times. At some stage (during some cycle) of this process the compressive stresses can exceed the yield limit of the oceanic lithosphere. The result will be a decrease in the buckling wavelength to the value of 200-300 km (see Section 3.8), a localization of deformation, and formation of a new subduction zone (as was observed in experiments 3.4 and 3.5 in Figures 3.5 and 3.6). A scenario described appears to be similar to the situation associated with the India-Eurasia collision. Intraplate deformation in the Indian Ocean caused by this collision had almost ceased several million years ago (see Section 3.4.1), but thrusting in the Himalaya is active all this time. Judging by the high seismic activity in the intraplate deformation zone, deformation in the Indian Ocean is currently being reactivated. Therefore one can suppose that the thrusting in Himalaya ceases (or ceased) and the level of lithospheric compression is high. A high compression may result in deformation in other adjoining zones (for example, in Tien Shan, Pamir, Hindu-Kush) and/or in strike-slip faulting within the Asian Plate (in Tibet, for example, see [*Avouac and Tapponnier*, 1993]) affected by the collision. The next stage may be either the formation of a new major crustal thrust in the Southern Himalaya or the formation of a new subduction zone in the Indian Ocean as described in Section 3.4.1.

REFERENCES

Alvarez, W., Geological evidence for the geographical patterns of mantle return flow and driving mechanism of plate tectonics, *J.Geophys.Res., 87*, 6697-6710, 1982.

Angelier, J., N.Lyberis, X.Le Pichon, E.Barrier, and P.Huchon, The tectonic development of the Hellenic Arc and the Sea of Crete: a synthesis, *Tectonophysics, 86*, 159-196, 1982.

Artyushkov, E.V., *Geodynamics* (in Russian), 320pp., Nauka, Moscow, 1979.

Ashby, M.F., and R.A.Verrall, Micro mechanisms of flow and fracture and their relevance to the rheology of the upper mantle, *Phil. Trans. Roy. Soc. London, A288, No 1350*, 59-95, 1978.

Aubouin, T., et al., Leg84 of the Deep Sea Drilling Project: Subduction without accretion: Middle America Trench off Guatemala, *Nature, 297*, 458-460, 1982.

Avouac, J.P., and P. Tapponnier, Kinematic model of Active deformation in Central Asia, Geoph. Res. Lett., 20, 895-898, 1993.

Balakina, L.M., Earthquakes and the tectonics of the Pacific rim (in Russian), *Priroda, 3*, 27-33, 1984.

Baranov, B.V., and L.I.Lobkovsky, Shallow seismicity behind the Kuril arc and its relation to the Benioff zone (in Russian), *Dokl. Akad. Nauk SSSR, 255*, 67-71, 1980.

Barker, P.F., and I.A.Hill, Back-arc extension in the Scotia Sea, *Phil. Trans. Roy. Soc. London, No.A300*, 249-262, 1981.

Barrier, E., Un grand accident actif: la faille de la valee Longitudinale de Taiwan, *Rev. Geol. Dyn. Geogr. Phys., 26*, 43-58. 1985.

Barrier, E., and J.Angelier, Active collision in eastern Taiwan, Coastal Range, *Mem. Geol. Soc. China, 7*, 135-155, 1986.

Biju-Duval, B., and J.Dercourt, Les basins de la Mediterranec orientale representent - ils les restes d'un domaine oceanique, la Mesogee, onvert an Mesozoique et distinct de la Tethys?, *Soc. Geol. France Bull. 22*, 43-60, 1980.

Biq, C. C., Transcurrent buckling, transform faulting and transpression, their relevance in eastern Taiwan kinematics, *Petro. Geol. Taiwan, 20*, 1-39, 1972.

Bowin, C., R.S.Lu, C. S.Lee, and H.Schouten, Plate convergence and accretion in Taiwan-Luzon region, *AAPG, 62*, 1645-1672, 1978.

Bridgman, P.W., *Dimensional Analysis*, 2nd ed., 113pp., Yale Univ. Press, New Haven, Conn., 1931.

Buckingham, E., On physically similar systems; illustrations of the use of dimensional equations, *Phys. Rev.,4*, 345-376, 1914.

Carlson, R.L., and P.J.Melia, Subduction hinge migration, *Tectonophysics, 102*, 399-411, 1984.

Carlson, R.L., and Mortera-Gutierrez, Subduction hinge migration along the Izu-Bonin-Mariana Arc, *Tectonophysics, 181*, 331-344, 1990.

Carr, M.J., R.E.Stoiber, and C.L.Drake, The segmented nature of some continental margins, in *The Geology of Continental Margins*, edited by Burk, C.A., and C.L.Drake, N.Y., p. 105-114, 1974.

Chai, B.H., Structure and tectonic evolution of Taiwan, *Am. J. Sci., 272*, 389-432, 1972.

Chang, L.S., and W.R.Chi, Neogene Nannoplankton biostratigraphy in Taiwan and the tectonic implications, *Petroleum Geol. Taiwan, 19*, 93-141, 1983.

Chase, C.G., Extension behind island arcs and motions relative to hot-spots, *J. Geophys. Res., 83*, 5385-5387, 1978.

Cherepanov G.P., and L.V.Ershov, *Mechanics of Failure* (in Russian), 223pp., Mashinostroenie, Moscow, 1977.

Chi, W. R., J.Namson, and J.Suppe, Stratigraphic record of plate interactions in the Coastal Range of eastern Taiwan, *Mem. Geol. Soc. China, 4*, 155-194, 1981.

Cloetingh, S., and R.Wortel, Regional stress field of the Indian plate, *Geophys. Res. Lett., 12*, 79-80, 1985.

Cloetingh, S., R.Wortel, and N.J.Vlaar, On the initiation of subduction zone, *Pageoph., 129*, 7-25, 1989.

Dalziel, I.W.D., Back-arc extension in the southern Andes: A review and critical reappraisal, *Philos. Trans. R. Soc. London, Ser. A300*, 319-335, 1981.

Davy, Ph., and P.R.Cobbold, Experiments on shortening of a 4-layer model of the continental lithosphere, *Tectonophysics, 188*, 1-25, 1991.

Dewey, J.F., Episodicity, sequence and style at convergent plate boundaries, in *The Continental Crust and Its Mineral Resources*, edited by D.Strangway, Geol. Assoc. Canada, Toronto, Ont., Spec. Pep. 20, p. 553-574, 1980.

Dewey, J.F. and J.M.Bird, Mountain belts and the new global tectonics, *J. Geophys. Res., 75*, 2625-2647, 1970.

Dewey, J., and C.A.M.Sengor, Aegean and surrounding regions: complex multiplate and continuum tectonics in a convergent zone, *Geol. Soc. Am. Bull., 90*, 84-92, 1979.

De Wit, M.J., The evolution of the Scotia arc as a key to the reconstruction of southwestern Gondwanaland, *Tectonophysics, 37*, 53-81, 1977.

Dickinson, W.R., and D.R.Seely, Structure and stratigraphy of forearc regions, *Amer. Assoc. Petrol. Geol. Bull., 63*, 2-31, 1979.

Dorsey, R. J., Collapse of the Luzon volcanic arc during onset of arc continent collision: evidence from a Miocene-Pliocene unconformity, eastern Taiwan, *Tectonics*, 11, 177-191, 1992.

Dubinin, E.P., *Transform Faults of the Oceanic Lithosphere* (in Russian), 180pp., Mosk. Univ., Moscow, 1987.

Elsasser, W.M., Sea-floor spreading as thermal convection, *J. Geophys. Res., 76*, 1101-1112, 1971.

England, P., and R.Wortel, Some consequences of the subduction of young slabs, *Earth Planet. Sci. Lett., 47*, 403-415, 1980.

Fedorchenko, V.I., and B.N.Piskunov, Dynamic regime of the Kuril islands in the relation of island arcs nature, in *Tectonics of the Sea Bottom, Island Arcs and Continental Margins* (in Russian), Vladivostok, p. 75-81, 1977.

Fisher, R.L., and C.C.Engel, Ultramafic and basaltic rocks dredged from the neashore flank of the Tonga trench, *Geol. Soc. Am. Bull., 80*, 1373-1378, 1969.

Forsyth, D.W., Fault plane solutions and tectonics of the South Atlantic and Scotia Sea, *J. Geophys. Res., 80*, 1429-1443, 1975.

Forsyth, D., and S.Uyeda, On the relative importance of the driving forces of plate motion, *Geophys. J. R. Astr. Soc., 43*, 163-200, 1975.

Frank, F.G., Curvature of island arcs, *Nature, 220*, 363, 1968.

Frankel, A., and W.McCann, Moderate and large earthquakes in the South Sandwich arc: indicators of tectonic variation along a subduction zone, *J. Geophys. Res., 84*, 5571-5577, 1979.

Froidevaux, C., and B.L.Isacks, The mechanical state of the lithosphere in the Altiplano-Puna segment of the Andes, *Earth Planet. Sci. Lett., 71*, 305-314, 1984.

Froidevaux, C., S.Uyeda, and M.Uyeshima, Island arc tectonics, *Tectonophysics, 148*, 1-9, 1988.

Furukawa, Y., Depth of the decoupling plate interface and thermal structure under arcs, *J. Geophys. Res., 98*, 20005-20013, 1993.

Galushkin, Yu.I., and S.A.Ushakov, The global picture of instantaneous kinematics of lithospheric plates (in Russian), *Vestnik Mosk. Univ., No 4*, 20-31, 1978.

Gershuni G.Z., and E.M.Zhukhovitsky, *Convective Stability of Incompressible Fluids* (in Russian), 390pp., Nauka, Moscow, 1972.

Gnibidenko, G.S. et al., *Tectonics of the Kuril-Kamchatka Deep Sea Trench* (in Russian), 190pp., Nauka, Moscow, 1980.

Gnibidenko, H.S., G.I.Anosov, V.V.Argentov, and I.K.Pushkin, Tectonics of the Tonga-Kermadec trench and Ozbourn seamount junction area, *Tectonophysics, 112*, 357-383, 1985.

Goetze, C., and B.Evans, Stress and temperature in bending lithosphere as constrained by experimental rock mechanics, *Geophys. J. R. Astron. Soc., 59*, 463-478, 1979.

Grocholsky, A.L., and A.I.Shemenda, Modeling the subduction of seamounts under island arcs (in Russian), in *Life of the Earth*, p. 13-18, Mosk. Univ., Moscow, 1985.

Grow, J.A., Crustal and upper mantle structure of the Central Aleutian arc, *Geol. Soc. Amer. Bull., 84*, 2169-2191, 1973.

Gunn, R., Quantitative aspects of juxtaposed ocean deep, mountain chains and volcanic ranges, *Geophysics, 12*, 238- 251, 1947.

Gzovsky, M.V., *Principles of Tectonophysics* (in Russian), 445pp., Nauka, Moscow, 1975.

Hager, B.H., and R.J.O'Connel, Subduction zone dip angles and flow driven by plate motion, *Tectonophysics, 50*, 111- 113, 1978.

Hill, R., *The Mathematical Theory of Plasticity*, 407pp., Clarendon, Oxford, 1950.

Hill, I.A., and P.F.Barker, Evidence for Miocene back-arc spreading in the central Scotia Sea, *Geophys. J. Roy. Astron. Soc., 63*, 427-440, 1980.

Ho, C. S., A synthesis of the geologic evolution of Taiwan, *Tectonophysics, 144*, 159-180, 1986.

Ho, C. S., *An introduction to the geology of Taiwan: explanatory text of the geologic map of Taiwan* (2nd ed.), Ministry of Economic Affairs, R. O. C., 164pp, 1988.

Hsu, T.L., Neotectonics of the Longitudinal Valley of Taiwan, *Bull. Geol. Survey Taiwan, 25*, 53-62, 1976.

Hsui, A.T., Application of fluid mechanics principles to the study of geodynamic processes at trench-arc-back arc systems, *Pageoph, 128*, 661-681, 1988.

Hsui, A., and N.Toksoz, The evolution of thermal structure beneath a subduction zone, *Tectonophysics, 60*, 43-60, 1979.

Huang, C. Y., C. T. Shyu, S.B.Lin, T.Q.Lee, and D.D.Sheu, Marine geology in the arc-continent collision zone off southeastern Taiwan: Implications for Late Neogene evolution of the Coastal Range, *Marine Geology, 107*, 183-212, 1992.

Hubbert, M.K., Theory of scale models as applied to the study of geologic structures, *Bull. Geol. Soc. Am., 48*, 1459-1520, 1937.

Hussong, D.H., and S.Uyeda, Tectonic processes and the history of the Mariana arc, *DSDP, Leg. 60, N 54*, 909-929, 1982.

Huzita, K., Role of the Median Tectonic Line in the Quaternary tectonics of the Japanese Islands, *Mem. Geol. Soc. Jpn.*, 18: 129-153, 1980.

Jacoby, W.R., Paraffin model experiment of plate tectonics, *Tectonophysics, 35*, 103-113, 1976.

Jacoby, W.R., and H.Schmeling, Convection experiments and driving mechanism, *Geol. Rundsch., 70,* 207-230, 1981.

Jurdy, D.M., and M.Stefanick, Flow models for back-arc spreading, *Tectonophysics, 99,* 191-206, 1983.

Kachanov, L.M., *Principles of Plasticity Theory* (in Russian), 418pp., Nauka, Moscow, 1969.

Kanamori, H., Great earthquakes at island arcs and the lithosphere, *Tectonophysics, 12,* 187-198, 1971.

Karig, D.E., Origin and development of the marginal basins in the western Pacific, *J. Geophys. Res., 76,* 2542-2561, 1971.

Karig, D.E., Remnant arcs, *Geol. Soc. Am. Bull., 83,* 1057-1067, 1972.

Karig, D.E., Plate convergence between the Philippine and the Ryukyu Islands, *Marine Geology, 14,* 153-168, 1973.

Karig, D.E., and F.Sharman, Subduction and accretion in trenches, *Geol. Soc. Amer. Bull., 86,* 377-485, 1975.

Kelleher, J., and W.McCann, Buoyant zones, great earthquakes and unstable boundaries of subduction, *J. Geophys. Res., 81,* 4885-4896, 1976.

Kerchman, V.I., F.M.Erlikhman, L.I.Lobkovsky, and O.G.Sorokhtin, Elastico-plastic analysis of plate interaction during subduction, in *Geology of Seas and Oceans, Abstr. of the 7th All-union school of marine geology, 2* (in Russian), pp. 36-37, Institut Okeanologii Akad. Nauk SSSR, Moscow, 1986.

Khain, V.E., *Regional Tectonics* (in Russian), 547pp., Nedra, Moscow, 1971.

Kincaid, C., and P.Olson, An experimental study of subducting slab migration, *J. Geophys. Res., 92,* 13,831-13,840, 1987.

Kirby, S.H., Tectonic stress in the lithosphere: constraints provided by the experimental deformation of rocks, *J. Geophys. Res., 85,* 6353-6363, 1980.

Kirby, S.H., Rheology of the lithosphere, *Rev. Geophys., 21,* 1458-1487, 1983.

Kline, S.J., *Similitude and Approximation Theory,* 229pp., McGraw-Hill, New York, N.Y., 1965.

Kogan, M.G., Gravity field of the Kuril-Kamchatka arc and its relation to the thermal regime of the lithosphere, *J. Geophys. Res., 80,* 1381-1396, 1975.

Koltipin, A.V., *Paleodynamic Analysis of the Kotjak Range* (in Russian), 215pp., Mosk. Univ., Moscow, 1991.

Kovalev, A.A., *Mobilism and Geologic Criteria in Exploration* (in Russian), 220pp., Nedra, Moscow, 1985.

Lallemand, S., H.Okada, K.Otsuka, and L.Labeyrie, Compressive tectonics along the eastern margin of the Japan sea: evidence for eastward thrust, *Comptes Rendus Acad. Sci. Paris, 301,* Ser. II, (3), 201-206, 1985.

Langhaar, H.L., *Dimensional Analysis and Theory of Models,* 166pp., Wiley, New York, N.Y., 1951.

Lee, C., G.Jr.Shor, L.D.Bibee, R.S.Lu, and T.W.C.Hilde, Okinawa Trough: Origin of a back arc basin, *Marine Geology, 35,* 219-241, 1980.

Le Pichon, X., J.Bonnin, and G.Pautot, The Gibraltar end of the Azores-Gibraltar plate boundary: An example of compressive tectonics, paper presented at Upper Mantle Committee Symposium, Flagstaff, Ariz., July 18-26, 1970.

Letouzey, J., and M.Kimura, The Okinawa Trough: genesis of a back-arc basin developing along a continental margin, *Tectonophysics, 125,* 209-230, 1986.

Levchenko, O.V., Geological structure of a zone of intra plate deformation in the Central Indian Basin (in Russian), Ph.D. thesis, Institut Okeanologii Acad. Nauk SSSR, Moscow, 1986.

Levchenko, O.V., L.R.Merklin, and V.E.Milanovsky, On the structure of sedimentary cover and tectonics of a zone of intraplate deformation in the lithosphere of the Indian Ocean (in Russian), in *Geology of*

Oceans and Seas, Abstracts of the 9th All Union School of Marine Geology, vol.2, p. 56-57, Institut Okeanologii Akad. Nauk SSSR, Moscow, 1990.

Lewis, S.D., and D.E.Hayes, The tectonics of northward propagating subduction along eastern Luzon, Philippine Islands, *Geophys. Monogr., 27*, 56-78, 1983.

Lin, M.T., and Y.B.Tsai, Seismotectonics in the Taiwan-Luzon area, *Bull. Inst. Earth Sci., Acad. Sin.,1*, 51-82, 1981.

Liu, C.C., and S.B.Yu, Vertical crustal movements in eastern Taiwan and their tectonic implications, *Tectonophysics*, 183, 111-119, 1990.

Liu, C. S., S.Y.Liu, B.Y.Kuo, N.Lundberg, and D.Reed, Characteristics of the gravity and magnetic anomalies off southern Taiwan, *Acta Geologica Taiwanica*, Science Rep. of NTU, No.30, 123-130, 1992.

Lobkovsky, L.I., *Geodynamics of Spreading and Subduction Zones and a Two-Layered Plate Tectonics* (in Russian), 252pp., Nauka, Moscow, 1988.

Lobkovsky, L.I., and A.I.Shemenda, An analysis of plastic deformation and stresses in a lithosphere plates in a subduction zone based on physical modeling (in Russian), in *Problems of Theoretical Geodynamics and Tectonics of Lithospheric Plates*, p. 146-153, Institut Okeanologii Akad. Nauk SSSR, Moscow, 1981.

Lobkovsky, L.I., and O.G.Sorokhtin, Plastic deformation of oceanic lithosphere in a subduction zone (in Russian), in *Tectonics of Lithospheric Plates*, p. 22-52, Institut Okeanologii Akad. Nauk SSSR, Moscow, 1976.

Lobkovsky, L.I., O.G.Sorokhtin, and A.I.Shemenda, Modeling of island arc deformation leading to the formation of tectonic terraces and the occurrence of tsunamigenic earthquakes (in Russian), *Dokl. Akad. Nauk SSSR, 255*, 74-77, 1980.

Lu, C. Y., and K.J.Hsu, Tectonic evolution of the Taiwan mountain belt, *Petroleum Geology of Taiwan, 27*, 21-46, 1992.

Ludwig, W., and P.D.Rabinowitz, The collizion complex of the North Scotia Ridge, *J. Geophys. Res., 87*, 3731-3790, 1982.

Lundberg, N., D.L.Reed, and C.H.Liu, The submarine propagating tip of the Taiwan collision: shallow crustal structure and orogenic sedimentation, *Taicrust Workshop Proc.*, Taipei, Taiwan, 93-102, 1991.

Makris, J., A dynamic model of the Hellenic arc deduced from geophysical data, *Tectonophysics, 36*, 339-346, 1976.

Malkin, B.V., and A.I.Shemenda, Mechanism of rifting:considerations based on results of physical modeling and on geological and geophysical data, *Tectonophysics, 199*, 193-210, 1991.

Mattauer, M., Intracontinental subduction, crust-mantle decollement and crustal-stacking wedge in the Himalayas and other collision belts, *Collision Tectonics*, Geol. Soc. Special Publication, No. 19, 37-50, 1986.

McAdoo, D.C., and D.T.Sandwell, Folding of oceanic lithosphere, *J. Geophys. Res., 90*, 8563-8569, 1985.

McKenzie, D.E., Speculations on the consequences and causes of plate motions, *Geophys. J. Roy. Astron. Soc., 18*, 1-32, 1969.

McKenzie, D., Active plate tectonics of the Mediterranean region, *Geophys. J. Roy. Astron. Soc., 30*, 109-185, 1972.

McKenzie, D., Active tectonics of the Alpine-Hemalayan belt: the Aegean Sea and surrounding regions, *Geophys. J. Roy. Astron. Soc., 55*, 217-254, 1978.

Mercier, J. L'Arc Eggen, une bordure deformee de la plaque eurasiatique: reflexions sur un example d'etude neotectonique, *Soc. Geol. France Bull., 19*, 663-672, 1977.

Meulenkamp, J.E., M.J.R.Wortel, W.A.Van Wamel, W.Spakman, and E.Hoogerduyn Strating, On the

Hellenic subduction zone and the geodynamic evolution of Crete since the late Middle Miocene, *Tectonophysics, 146*, 203-215, 1988.

Minster, J.B., T.H.Jordan, P.Molnar, and E.Haines, Numerical modeling of instantaneous plate tectonics, *Geophys.J. R. Astron. Soc., 36*, 541-576, 1974.

Mitchell, A.H.G., Initiation of subduction of post-collision foreland thrusting and back-thrusting, *J. Geogynamics, 1*, 103-120, 1984.

Molnar, P., and T.Atwater, Interarc spreading and cordileran tectonics as alternates related to the age of subducted oceanic lithosphere, *Earth Planet. Sci. Lett., 41*, 330-340, 1978.

Monin, A.S., and O.G.Sorokhtin, A possible nature of tectonic cycles, in *Oceanology, Geophysics of the Ocean, v.2 Geodynamics* (in Russian), p. 119-130, Nauka, Moscow, 1979.

Moore, J.G., and D.E.Karig, Sedimentology, structural geology and tectonics of the Shikoku subduction zone, southwestern Japan, *Geol. Soc. Amer. Bull., 87*, 1259-1268, 1976.

Morelli, C., M.Pisani, and C.Ganter, Geophysical studies in the Aegean Sea and in the Eastern Mediterranean, *Boll. Geoph. Teor. Appl., 66*, 127-167, 1975.

Muller, S., and W.Lowrie, Die geodynamische Entwicklung des westlichen Mittelmeerraums und der Alpen: Vermessung, *Photogrammetrie, Kulturtechnik, 12*, 470-495, 1980.

Murphy, R.W., The Manila Trench - west Taiwan foldbelt: a fliped subduction zone, *Geol. Soc. Malaysia Bull.,6*, p.27-42, 1973

Nakamura, K., Possible nascent trench along the eastern Japan Sea as the convergent boundary between Eurasian and North American plates (in Japanese with English abstract), *Bull. Earthq. Res. Ist., Univ. Tokyo, 58*, 711-722, 1983.

Nakamura, K., and S.Uyeda, Stress gradient in arc-back arc regions and plate subduction, *J. Geophys. Res., 85*, 6419- 6428, 1980.

Nelson, T.H., and P.G.Temple, Mainstream mantle convection: A geologic analysis of plate motion, *Amer. Assoc. Petrol. Geol. Bull., 56*, 226-246, 1972.

Nikolaevsky, V.N., Mechanics of geomaterials and earthquakes, in *Mechanics of Deformable Solids* (in Russian), p. 149-230, Vsesouzniy Institut Nauchno-Texnicheskoy Informatsii, Moscow, 1983.

Nishenko, S., and W.McCann, Large thrust earthquakes and tsunamis: implications for the development of forearc basins, *J. Geophys. Res., 84*, 573-584, 1979.

Ogawa, Y., T.Seno, H.Akiyoshi, H.Tokuyama, K.Fujioka, and H.Taniguchi, Structure and development of the Sagami trough and the Boso triple junction, *Tectonophysics, 160*, 135-150, 1989.

Oxburgh, E.R., and D.L.Turcotte, The physico-chemical behavior of the descending lithosphere, *Tectonophysics, 32*, 107-128, 1976.

Oxburgh, E.R., and E.M.Parmentier, Compositional and density stratificationin oceanic lithosphere - causes and conseguences, *J. Geol. Soc. London, 133*, 343-354, 1977.

Packham, G.H., and D.A.Falvey, An hypothesis for the formation of the marginal seas in the western Pacific, *Tectonophysics, 11*, 79-109, 1971.

Page, B. M., and J.Suppe, The Pliocene Lichi melange of Taiwan: its plate tectonic and olistostromal origin, *Am. J. Sci., 281*, 193-227, 1981.

Papazachos, B., Distribution of seismic foci in the Mediterranean and surrounding area and its tectonic implications, *Geophys. J. R. Astron. Soc., 33*, 421-430, 1973.

Papazachos, B., Seismotectonics of the northern Aegean area, *Tectonophysics, 33*, 199-209, 1976.

Pelayo, A.M., and A.Wiens, Seismotectonics and relative plate motions in the Scotia sea region, *J. Geoph. Res., 94*, 7293-7320, 1989.

Peterson, E.T., and T.Seno, Factors affecting seismic moment release in subduction zones, *J. Geophys.*

Res., 89, 10233-10248, 1984.

Prozorov, Yu.I., and A.I.Shemenda, Geodynamics of the Aegean region (in Russian), *Geotectonics, No 6*, 109-117, 1987.

Rabinowitz, P.D., Free-air gravity anomalies bordering the continental margin of Argentina, *Argentina map series*, Amer. Assoc. Petrol. Geol., Tulsa, Oklahoma, 1977.

Ramberg, H., *Gravity, Deformation and the Earth's Crust*, 2nd ed., 452pp., Acad. Press, London, 1981.

Ranalli, G., and D.Murphy, Rheologic stratification of the lithosphere, *Tectonophysics, 132*, 281-295, 1987.

Ranalli, G., *Rheology of the Earth*, 366pp., Allen and Unwin, Boston, 1987.

Rudnicki, J.W., and J.R.Rice, Conditions for the localization of deformation in pressure-sensitive, dilatant materials, *J. Mech. Phys. Solids, 23*, 371-394, 1975.

Ruff, L., and H.Kanamori, Seismicity and the subduction process, *Phys. Earth Planet. Inter. 23*, 240-252, 1980.

Sacks, I.S. The subduction of young lithosphere, *J. Geophys. Res., 34*, 3355-3366, 1983.

Sazhin, N.B., Gravity map of Africa, Moscow, MINGEO USSR, 1975.

Sedov, L.I., *Similarity and Dimensional Methods in Mechanics*, 363pp., Academic Press, London, 1959.

Sedov, L.I., *Similarity and Dimensional Analysis in Mechanics* (in Russian), 7th ed., 395pp., Nauka, Moskow, 1972.

Sedov, L.I., *Mechanics of Continua* (in Russian), v. 2, 570 pp., Nauka, Moscow, 1976.

Sengor, A., and Y.Yilmaz, Tethyan evolution of Turkey: a plate tectonic approach, *Tectonophysics, 75*, 181-241, 1981.

Seno, T., The instaneous rotation vector of the Philippine Sea plate relative to the Eurasian plate, *Tectonophysics, 42*, 209-226, 1977.

Shemenda, A.I., Physical modeling of a subduction zone of oceanic lithospheric plates (in Russian), *Dokl. Akad. Nauk SSSR, 246*, 1334-1337, 1979.

Shemenda, A.I., The conditions and techniques of physical modeling of the subduction of lithosphere plates, in *Problems of Theoretical Geodynamics and the Tectonics of Lithosphere Plates* (in Russian), p. 146-154, Institut Okeanologii Akad. Nauk SSSR, Moscow, 1981.

Shemenda, A.I., Similarity criteria in the mechanical modeling of tectonic processes, *Geol. Geofiz., 10*, 10-19, 1983.

Shemenda, A.I., The conditions and some results from physical modeling of lithospheric deformation, in *Experimentation and Modeling in Geological Researches* (in Russian), p. 27-46, Institut Geologii i Geofiziki Sibirskogo Otdeleniya Akad. Nauk SSSR, Novosibirsk, Russia, 1984.

Shemenda, A.I., Modeling the tectonic opening mechanism of certain types of marginal seas, *Oceanology, USSR Acad. of Sci., v.25*, No 2, 204-210, 1985.

Shemenda, A.I., On the spreading mechanism for marginal seas, *Oceanology, USSR Acad. of Sci., v.26*, No 1, 109-116, 1986.

Shemenda, A.I., Physical similarity problem in experimental geodynamics, in *Experimental Tectonics in Solving of theTheoretical and Applied Tasks of Geology* (in Russian), p. 14-16, Institut Geofiziki Ukrainskoy Akad. Nauk, Kiev, 1987.

Shemenda, A.I., Results from physical modeling of horizontal compression in the lithosphere (in Russian), *Dokl. Akad. Nauk SSSR, 307*, 345-350, 1989*a*.

Shemenda, A.I., Modeling of intraplate deformation in the northeastern Indian Ocean (in Russian), *Geotectonics, 3*, 37-49, 1989*b*.

Shemenda, A.I., Modeling of large-scale deformation in the lithosphere (in Russian), in *Experimental*

Tectonics, p. 252-262, Nauka, Moscow, 1989*c*.

Shemenda, A.I., Horizontal lithosphere compression and subduction: constraints provided by physical modeling, *J. Geophys. Res.*, *97*, 11097-11116, 1992.

Shemenda, A. I., Subduction of the lithosphere and back arc dynamics: insights from physical modeling, *J. Geophys. Res.*, *98*, 16167-16185, 1993.

Shemenda, A.I., and A.L.Grocholsky, Geodynamics of the South Antilles Region, *Geotectonics, 20*, 58-66, 1986.

Shemenda, A.I. and A.L.Grocholsky, A formation and evolution of overlapping spreading centers (constrained on the basis of physical modeling), *Tectonophysics, 199*, p. 398-404, 1991.

Shemenda, A.I., and A.L.Grocholsky, Physical modeling of lithosphere subduction in collision zones, *Tectonophysics, 216*, 273-290, 1992.

Shemenda, A.I., and A.L.Grocholsky, Physical modeling of slow seafloor spreading, *J. Geoph. Res.*, 99, 9137-9153, 1994.

Shemenda, A.I., C.-H.Hsieh, and R.K.Yang, Geodynamic model of the Taiwan area, *Proceeding of the Fourth Taiwan Symposium on Geophysics*, Taiwan, 525-535, 1992*a*.

Shemenda, A. I., C.-H.Hsieh, and R.K.Yang, Reversal subduction and a geodynamic model of the collision in Taiwan, *Acta Geologica Taiwanica,* Science Rep. of NTU, No.30, 101-104, 1992*b*.

Sherman S.I., Physical experiment in tectonics and similarity theories (in Russian), *Geol. Geofiz., No 3*, 8-18, 1984.

Shimamoto, T., Application of the Pi-theorem to the similarity criteria of slow deformation of inhomogeneous viscous fluid, *Tectonophysics, 22*, 253-263, 1974.

Sorokhtin O.G., *Global Evolution of the Earth* (in Russian), 184pp., Nauka, Moscow, 1974.

Stein, S., and E.A.Okal, Seismicity and tectonics of the 90°E Ridge area: Evidence for internal deformation of the Indian plate, *J. Geophys. Res., 83,* 2233-2246, 1978.

Stevenson, D.L., and J.S.Turner, Angle of subduction, *Nature, 270*, 334-336, 1977.

Studies in the East Asian, Tectonics and Resources, JDOE SEATAK CCOP JOC, 1980.

Sugi, N., K.Chinzei, and S.Uyeda, Vertical crustal moments of the notheast Japan since middle Miocene, in *Geodynamics of the Western Pacific-Indonesian Region, Geodynamics, Ser. 11*, edited by Hilde, T., and S. Uyeda, Amer. Geophys. Un./Geol. Soc. Amer., p. 317-329, 1983.

Suppe, J., Mechanics of mountain building and metamorphism in Taiwan, *Mem. Geol. Soc. China, 4*, 67-89, 1981.

Suppe, J., Kinematics of arc-continent collision, flipping of subduction, and back arc spreading near Taiwan, *Mem. Geol. Soc. China, 6*, 21-24, 1984.

Sykes, L.R., Seismicity of the Indian ocean and a possible nascent island arc between Ceylon and Australia, *J. Geophys. Res., 75,* 5041-5055, 1970.

Tada, A., The 1983 Japan Sea earthquake and crustal deformation, *The Earth Monthly, 6*: 18-23, 1984 (in Japanese).

Tamaki, K., Active features and tectonics along the eastern margine of the Japan Sea, *The Earth Monthly, 6*: 38-48, 1984 (in Japanese).

Tapponnier, P., Evolution tectonoque du systeme Alpin en Mediterranee: poinconnement et ecrasement rigide-plastique, *Soc. Geol. France Bull., 19*, 437-460, 1977.

Tarakanov, R.Z., and Ch.U.Kim, Seismicity in the Kuril-Kamchatka zone and identification of transcurrent faults, in *Seismicity and Earthquake Mechanisms in the Far East* (in Russian), p. 17-25, SahKNII, Sachalin, 1980.

Taylor, B., and G.D.Karner, On the evolution of marginal basins, *Rev. Geophys. and Space Phys., 21*,

1727-1741, 1983.

Teng, L. S., Geotectonic evolution of late Cenozoic Penglai collision in Taiwan, *Tectonophysics, 183*, 57-76, 1990.

Teng, L. S., Geotectonic evolution of Tertiary continental margin basins of Taiwan, *Petrol. Geol. of Taiwan, 27*, 1-19,1992.

Tishchenko, E.A., A geodynamic analysis of deformation in lithospheric plates in the areas of compression by numerical modeling (in Russian), Ph.D. thesis, Institut Geologii i Geofiziki Sibirskogo Otdeleniya Akad. Nauk SSSR, Novosibirsk, Russia, 1985.

Toksoz, M.N., N.H.Sleep, and A.T.Smith, Evolution of the downgoing lithosphere and the mechanism of deep focus earthquakes, *Geophys. J. Roy. Astron. Soc., 35*, 285-310, 1973.

Tovish, A., and G.Schubert, Island arc curvature, velocity of convergence and angle of subduction, *Geophys. Res. Lett., 5*, 329-332, 1978.

Tullis, T., Evidence that lithosphere plates act as anchors (abstr.), *EOS Trans. AGU, 53*, 522, 1972.

Turcotte, D.L., J.Y.Liu, and F.H.Kulhawy, The role of an intra crustal asthenosphere on the behavior of major strike-slip faults, *J. Geophys. Res., 89*, 5801-5816, 1984.

Turcotte, D.L., and G. Schubert, *Geodynamics: Applications of Continuum Physics to Geological Problems*, pp. 337-346, John Wiley, New York, 1982.

Turner, J.S., Convection in the mantle: A laboratory model with temperature-dependent viscosity, *Earth Planet. Sci. Lett., 17*, 369-374, 1973.

Ushakov, S.A., Viscosity and dynamic processes in the crust and upper mantle, *Vestn. Mosk. Univ., Ser.4, Geol., 23*(1), 33-44, 1968.

Ushakov, S.A., et al., *Gravity Field and Sea Floor Topography* (in Russian), 296pp., Nedra, St.Petersburg, Russia, 1979.

Uyeda, S., Subduction zones: an introduction to comparative subductology, *Tectonophysics, 81*, 133-159, 1982.

Uyeda, S., Subduction zones: Their diversity, mechanism and human impacts, *GeoJournal 8.4*, 381-406, 1984.

Uyeda, S., and H.Kanamori,. Back arc opening and the mode of subduction, *J. Geophys. Res., 84*, 1049-1061, 1979.

Van der Held, E.F.M., Currents inside the Earth I, Proc. Kon. Ned. Acad., *Van Western-Schappen, Ser. B 72*, 256-286, Amsterdam, 1969.

Verzhbitsky, E.V., et al., New evidence on the geologic structure and tectonic evolution of the Gorringe seamount chain, *Geotectonics, 1*, 12-22, 1989.

Vogt, P.R., A.Lowrie, D.R.Bracey, and R.N.Hey, Subduction of aseismic ocean ridges: Effects on shape, seismicity and other characteristics of consuming plate boundaries, *Spec. Pap. Geol. Soc. Am., 172*, 59pp., 1976.

von Huene, R., and S.Uyeda, A summary of results from the IPOD active margin transacts across the Japan, Mariana and Mid-American convergent margins, *Oceanol. Asta (Suppl.), Colloque C3*, 26th Geol. Congr., Paris, 233-239, 1981.

von Huene, R., M.Langseth, N.Nasu, and H.Okada, Summury, Japan trench transact. *Init. Rep. DSDP LVI-LVII, pt.1*, US Dovern. Print. Off., Washington, DC, 473-488, 1980.

Watts, A.B., and M.Talwani, Gravity anomalies seaward of deep-ocean trenches and their tectonic implications, *Geophys. J. R. Astron. Soc., 36*, 57-70, 1974.

Watts, A.B., and M.Talwani, Gravity effect of downgoing lithospheric slabs beneath island arcs, *Geol. Soc. Am. Bull., 86*, 1-4, 1975.

Weijermars, R., and H.Schmeling, Scaling of newtonian and non newtonian fluid dynamics without inertia for quantitative modeling of rock flow due to gravity (including the concept of rheologic similarity), *Phys. Earth Planet. Inter., 43*, 316-330, 1986.

Weissel, J., R.Anderson, and C.Geller, Deformation of the Indo-Australian plate, *Nature, 287*, 284-291, 1980.

Whitmarsh, R.B., The Owen basin off the south-east margin of Arabia and evolution of the Owen fracture zone, *Geophys. J. R. Astron. Soc., 58*, 441-470, 1979.

Wiens, D.A., S.Stein, C.Demets, R.G.Gordon, and C.Stain, Plate tectonic model for Indian ocean "intraplate" deformation, *Tectonophysics, 132*, 37-48, 1986.

Wu, F.T., Recent tectonics of Taiwan, *J. Phys. Earth., 26, Supp.*, S265-S299, 1978.

Yamano, M., and S.Uyeda, Possible effects of collision on plate motions, *Tectonophysics, 119*, 223-244, 1985.

Yamaoka, K., Y.Fukao, and M.Kumazawa, Spherical shell tectonics: Effects of sphericity and inextensibility on the geometry of the descending lithosphere, *Rev. of Geophys,. 24*, 27-53, 1986.

Yeh, Y.H., and H.Y.Yen, Gravity anomalies of Taiwan and their tectonic implication, *Taicrust Workshop Proc.*, Taipei, Taiwan, 175-184, 1991.

Yokokura, T., On subduction dip angles, *Tectonophysics, 77*, 63-77, 1981.

Yoshii, T., Upper mantle structure beneath the north Pacific and the marginal seas, *J. Phys. Earth, 21*, 313-328, 1973.

Yu, S.B. and Y.B.Tsai, A study of microseismicity and Crustal deformation of the Kuang Fu-Fuli Area in Eastern Taiwan, *Bull. Inst. Earth Sci., Academia Sinica, 2*, 1-17, 1982.

Yu, S.B., D.D. Jackson, G.K.Yu, and C.C.Liu, Dislocation model for crustal deformation in trhe Longitudinal Valley area, eastern Taiwan, *Tectonophysics, 183*, 97-109, 1990.

Yuen, D.A., and G.Schubert, Asthenospheric shear flow: thermally stable or unstable?, *Geophys. Res. Lett., 4*, 503-506, 1977.

Zonenshain, L.P., and L.A.Savostin, *Introduction into Geodynamics* (in Russian), 290pp., Nedra, Moscow, 1979.

Zuber, M.T., Compression of oceanic lithosphere: An analysis of intraplate deformation in the Central Indian Basin, *J. Geophys. Res., 92*, 4817-4825, 1987.

NOTATIONS

C_p	specific heat.
d	width of the marginal basin and the volcanic arc.
e_{ij}	deviatoric strain rate tensor.
E	lithospheric modulus of elasticity.
F	horizontal compressive force of the overriding plate.
F_f	horizontal component of friction force between the plates in the subduction zone.
F_t	horizontal nonhydrostatic tensional force of the overriding plate.
g	acceleration of gravity.
Δg	free air gravity anomaly.
G	gravitational constant.
h	thickness of thinned segment of the plate and plastic hardening modulus.
h_m	maximum thickness of thinned segment of the plate under which its failure occurs.
h_s, d_s	thickness and length of the sedimentary wedge on the passive continental margin.
Δh_n	nonisostatic vertical displacement of the surface of the overriding plate.
H	lithospheric thickness.
H_c	crustal thickness.
H_l	thickness of the oceanic lithosphere.
H_m	thickness of the mantle layer of the continental lithosphere.
H_v	thickness of the crust in the volcanic arc.
$H_{\dot{e}}$	intensity of shear strain rate.
K_c	critical coefficient of stress intensity.
Δl	displacement.
L	latent heat.
p	pressure.
P	hydrostatic pressure.
P_n	contact pressure between plates.
P_0	tensional nonhydrostatic stress in the overriding plate due to its tendency to spread into the trench.
ΔP	excess pressure in the asthenosphere.

s_{ij} deviatoric stress tensor

t time.

T temperature.

T_s, T_l temperatures of solidus and liquidus.

V velocity.

w deflection of the lithosphere surface from the isostatic equilibrium level.

x horizontal coordinate.

z vertical coordinate.

z_m maximum depth for opening of vertical cut in the plate.

z_o trench depth with respect to the level of hydrostatic equilibrium of lithosphere surface.

β, α families of slip-lines.

β dilatancy factor.

β_v coefficient of thermal expansion.

δ_{ij} Kronecker delta.

ε strain.

$\dot{\varepsilon}$ strain rate.

ε_{ij} strain tensor.

$\dot{\varepsilon}_{ij}$ strain rate tensor.

$\Delta\varepsilon_p$ plastic strain prior to failure.

ζ depth of the surface of the undisturbed oceanic plate to that of the continental lithosphere.

η asthenosphere viscosity.

κ thermal diffusivity.

λ wavelength of lithosphere flexure and thermal conductivity.

μ internal friction coefficient.

ν Poisson's ratio.

ρ density.

ρ_a asthenospheric density.

ρ_c crustal density.

ρ_l density of oceanic lithosphere.

ρ_o overriding plate density.

ρ_v density of the volcanic arc.

$\Delta\rho$	defined as $\rho_l - \rho_a$.
σ_h	horizontal nonhydrostatic stress in the lithosphere.
σ_n	nonhydrostatic pressure between plates in subduction zone.
σ_s	lithospheric yield limit under normal loading.
σ_v	vertical nonhydrostatic pressure in the interplate surface.
σ_a	strength of the volcanic arc.
σ_o	mean pressure.
σ_1, σ_2	principal stresses.
σ_{ij}	stress tensor.
τ	shear stress.
$\bar{\tau}$	intensity of deviatoric stresses.
τ_a	friction at the base of lithosphere.
τ_n	friction between plates.
τ_s	lithospheric shear yield limit.
τ_{ij}	shear stress tensor.
$\Delta\tau_p$	stress drop due to failure in the material.
ψ	dip angle of the interplate surface in the subduction zone.
ω, ξ	parameters characterizing specific slip-lines.

INDEX

MODERN APPROACHES IN GEOPHYSICS

formerly *Seismology and Exploration Geophysics*

1. E.I. Galperin: *Vertical Seismic Profiling and Its Exploration Potential.* 1985 ISBN 90-277-1450-9
2. E.I. Galperin, I.L. Nersesov and R.M. Galperina: *Borehole Seismology.* 1986 ISBN 90-277-1967-5
3. Jean-Pierre Cordier: *Velocities in Reflection Seismology.* 1985 ISBN 90-277-2024-X
4. Gregg Parkes and Les Hatton: *The Marine Seismic Source.* 1986 ISBN 90-277-2228-5
5. Guust Nolet (ed.): *Seismic Tomography.* 1987 ISBN 90-277-2521-7
6. N.J. Vlaar, G. Nolet, M.J.R. Wortel and S.A.P.L. Cloetingh (eds.): *Mathematical Geophysics.* 1988 ISBN 90-277-2620-5
7. J. Bonnin, M. Cara, A. Cisternas and R. Fantechi (eds.): *Seismic Hazard in Mediterranean Regions.* 1988 ISBN 90-277-2779-1
8. Paul L. Stoffa (ed.): *Tau-p: A Plane Wave Approach to the Analysis of Seismic Data.* 1989 ISBN 0-7923-0038-6
9. V.I. Keilis-Borok (ed.): *Seismic Surface Waves in a Laterally Inhomogeneous Earth.* 1989 ISBN 0-7923-0044-0
10. V. Babuska and M. Cara: *Seismic Anisotropy in the Earth.* 1991 ISBN 0-7923-1321-6
11. A.I. Shemenda: *Subduction.* Insights from Physical Modeling. 1994 ISBN 0-7923-3042-0

KLUWER ACADEMIC PUBLISHERS – DORDRECHT / BOSTON / LONDON

MIX
Papier aus verantwortungsvollen Quellen
Paper from responsible sources
FSC® C105338
FSC
www.fsc.org

If you have any concerns about our products,
you can contact us on
ProductSafety@springernature.com

In case Publisher is established outside the EU,
the EU authorized representative is:
Springer Nature Customer Service Center GmbH
Europaplatz 3, 69115 Heidelberg, Germany

Printed by Libri Plureos GmbH
in Hamburg, Germany